Electromagnetism

Electromagnetism

by JOHN C. SLATER, Ph.D.

Graduate Research Professor, University of Florida

Institute Professor Emeritus
Massachusetts Institute of Technology

and NATHANIEL H. FRANK, Sc.D.

Professor of Physics, Emeritus
Massachusetts Institute of Technology

DOVER PUBLICATIONS, INC.

New York

Published in Canada by General Publishing Company, Ltd., 30 Lesmill Road, Don Mills, Toronto, Ontario.

This Dover edition, first published in 1969, is an unabridged and slightly corrected republication of the work originally published in 1947 by the Mc-Graw-Hill Book Company, Inc.

Standard Book Number: 486-62263-0
Library of Congress Catalog Card Number: 69-17476

Manufactured in the United States of America
Dover Publications, Inc.
180 Varick Street
New York, N. Y. 10014

PREFACE

The present book is the second of several volumes which are intended to replace the *Introduction to Theoretical Physics* written by the same authors in 1933. By separating the material on mechanics, on electromagnetism, and on the quantum theory, we believe that it is possible to give a somewhat better rounded treatment of each of these fields, which will be more useful to the teacher and the student. We have taken advantage of the opportunity to give a considerably more complete treatment of the foundations of electrostatics and magneto-statics, and to introduce some of the new developments in electromagnetic theory since 1933. At the same time, we have tried to preserve, and even to extend, the general unity of treatment which we believed so important when the earlier book was published. We have a conviction that the teaching of theoretical physics in a number of separate courses, as in mechanics, electromagnetic theory, potential theory, thermodynamics, and modern physics, tends to keep a student from seeing the unity of physics, and from appreciating the importance of applying principles developed for one branch of science to the problems of another.

We have developed electromagnetism from first principles, and have included in the appendixes enough of the necessary mathematical background so that the student familiar with the calculus and differential equations can follow the work, without further training in mathematics or in mathematical physics. Nevertheless, electromagnetic theory is a later historical development than mechanics, and it makes use of many mathematical methods, as in vector and tensor analysis, potential theory, and partial differential equations, which were first developed for mechanics, and find their most straightforward applications in mechanical problems. Although we feel that the present volume is designed so that it can be used by itself, a thorough grounding in mechanics is a very desirable prerequisite. The student who first familiarizes himself with the companion volume on mechanics will find many similarities in treatment of the two fields, which will enhance his understanding of both.

Electromagnetic theory has developed in two principal directions since its original formulation in the middle of the last century: toward the electromagnetic theory of light, treating very short wave lengths, and providing the mathematical foundation for the theory of optics; and toward the longer wave lengths encountered in electrical engineering and radio.   In the last few years, the practical applications have come at constantly shorter wave lengths, until with the recent development of microwaves the gap between the two branches of the subject has been practically closed.   We handle problems of both types, making no distinction between the methods used for them.

In 1933, it was almost universal practice to use the Gaussian system of units for handling electromagnetic problems, and *Introduction to Theoretical Physics* is written in terms of those units.   Since then, the mks, or meter-kilogram-second, system has come into common use, and we have made use of that system in the present volume, believing that its advantages over the Gaussian system are considerable.   We give in an appendix, however, a discussion of the various systems of units, including a formulation of all the important equations in terms of the Gaussian system, so that the reader will not be at a disadvantage in consulting other books using the Gaussian units, and so that he will understand properly the relations among the various systems.

We quote from the preface of *Introduction to Theoretical Physics:*

"In a book of such wide scope, it is inevitable that many important subjects are treated in a cursory manner.   An effort has been made to present enough of the groundwork of each subject so that not only is further work facilitated, but also the position of these subjects in a more general scheme of physical thought is clearly shown.   In spite of this, however, the student will of course make much use of other references, and we give a list of references, by no means exhaustive, but suggesting a few titles in each field which a student who has mastered the material of this book should be able to appreciate.

"At the end of each chapter is a set of problems.   The ability to work problems, in our opinion, is essential to a proper understanding cf physics, and it is hoped that these problems will provide useful practice.   At the same time, in many cases, the problems have been used to extend and amplify the discussion of the subject matter, where limitations of space made such discussion impossible in the text.   The attempt has been made, though we are conscious of having fallen far short of succeeding in it, to carry each branch of the subject far enough so that definite calculations can be made with it.   Thus a far surer mastery is attained than in a merely descriptive discussion.

"Finally, we wish to remind the reader that the book is very definitely one on theoretical physics. Though at times descriptive material, and descriptions of experimental results, are included, it is in general assumed that the reader has a fair knowledge of experimental physics, of the grade generally covered in intermediate college courses. No doubt it is unfortunate, in view of the unity which we have stressed, to separate the theoretical side of the subject from the experimental in this way. This is particularly true when one remembers that the greatest difficulty which the student has in mastering theoretical physics comes in learning how to apply mathematics to a physical situation, how to formulate a problem mathematically, rather than in solving the problem when it is once formulated. We have tried wherever possible, in problems and text, to bridge the gap between pure mathematics and experimental physics. But the only satisfactory answer to this difficulty is a broad training in which theoretical physics goes side by side with experimental physics and practical laboratory work. The same ability to overcome obstacles, the same ingenuity in devising one method of procedure when another fails, the same physical intuition leading one to perceive the answer to a problem through a mass of intervening detail, the same critical judgment leading one to distinguish right from wrong procedures, and to appraise results carefully on the ground of physical plausibility, are required in theoretical and in experimental physics. Leaks in vacuum systems or in electric circuits have their counterparts in the many disastrous things that can happen to equations. And it is often as hard to devise a mathematical system to deal with a difficult problem, without unjustifiable approximations and impossible complications, as it is to design apparatus for measuring a difficult quantity or detecting a new effect. These things cannot be taught. They come only from that combination of inherent insight and faithful practice which is necessary to the successful physicist. But half the battle is over if the student approaches theoretical physics, not as a set of mysterious formulas, or as a dull routine to be learned, but as a collection of methods, of tools, of apparatus, subject to the same sort of rules as other physical apparatus, and yielding physical results of great importance. . . . The aim has constantly been, not to teach a great collection of facts, but to teach mastery of the tools by which the facts have been discovered and by which future discoveries will be made."

JOHN C. SLATER
NATHANIEL H. FRANK

CAMBRIDGE, MASS.,
July, 1947.

CONTENTS

CHAPTER IX

ELECTRON THEORY AND DISPERSION

CHAPTER X

REFLECTION AND REFRACTION OF ELECTROMAGNETIC WAVES

CHAPTER XI

WAVE GUIDES AND CAVITY RESONATORS

CHAPTER XII

SPHERICAL ELECTROMAGNETIC WAVES

CHAPTER XIII
HUYGENS' PRINCIPLE AND GREEN'S THEOREM

CHAPTER XIV
FRESNEL AND FRAUNHOFER DIFFRACTION

APPENDIX I

APPENDIX II

APPENDIX III

APPENDIX IV

APPENDIX V

APPENDIX VI

APPENDIX VII

CHAPTER I

THE FIELD THEORY OF ELECTROMAGNETISM

A dynamical problem has two aspects: mechanics, the determination of the accelerations and hence of the motions, once the forces are given; and the study of the forces acting under the existing circumstances. The basic principles of mechanics are simple. In its classical form, mechanics is based on Newton's laws of motion, laws discovered and formulated nearly three hundred years ago. The developments since then have been technical, mathematical improvements in the way of formulating the laws and solving the resulting mathematical problems, rather than additions to our fundamental knowledge of mechanics. Only in the present century, with wave mechanics, has there been a change in the underlying structure of the subject.

The study of forces, on the other hand, is difficult and complex. The first forces brought into mathematical formulation were gravitational forces, as seen in planetary motion. Next were elastic forces. Then followed electric and magnetic forces, which are the subject of this volume. Their study was mostly a product of the nineteenth century. During the present century, it has become clear that electromagnetic forces are of far wider application than was first supposed. It has become evident that, instead of being active only in electrostatic and magnetostatic experiments, and in electromagnetic applications such as the telegraph, dynamo, and radio, the forces between the nuclei and electrons of single atoms, the chemical forces between atoms and molecules, the forces of cohesion and elasticity holding solids together, are all of an electric nature. We might be tempted to generalize and suppose that all forces are electromagnetic, but this appears to be carrying things too far. The prevailing evidence at present indicates that the intranuclear forces, holding together the various fundamental particles of which the nucleus is composed, are not of electromagnetic origin. These forces, of enormous magnitude, and appearing in the phenomena of radioactivity and of nuclear fission, appear subject to laws somewhat analogous to the electromagnetic laws, but fundamentally different. In spite of this, the range of phenomena governed by electromagnetic theory is very wide, and it carries us rather far into the structure of matter, of electrons and nuclei and atoms and molecules, if we wish to understand it completely. The equations underlying the theory, Maxwell's equations, are relatively simple, but not nearly so simple as Newton's laws of motion. Instead of stating the whole fundamental formulation of the subject in the first chapter, as one can when dealing with mechanics, about half of the present book is taken up with a complete formulation of Maxwell's equations. We start with simple types of force, electrostatic and magnetostatic, and gradually work up to problems of electromagnetic induction and related subjects, all of which are formulated in Maxwell's equations.

In the development of electromagnetic theory, there has been a continual and significant trend, which in a way has set the pattern for the development of all of theoretical physics. This has been the trend away from the concept known as "action at a distance" toward the concept of field theory. The classical example of action at a distance is gravitation, in which simple nonrelativistic theory states that any two particles in the universe exert a gravitational force on each other, acting along the line joining them, proportional to the product of their masses, and inversely proportional to the square of the distance between them. Such a force, depending only on the positions of the particles, quite independent of any intervening objects, is simple to think about, and formed the basis of most of physical thought from the time of Newton, in the latter half of the seventeenth century, on well into the nineteenth century. The first electric and magnetic laws to be discovered fitted in well with the pattern. First among these was Coulomb's law. Coulomb investigated the forces between electrically charged objects, and found that the force between two such objects was in the line joining them, proportional to the product of their charges (which could be defined by an experiment based on this observation), and inversely proportional to the square of the distance between them, in striking analogy to the law of gravitation. Magnets similarly fell in with the scheme. A theory of the forces between permanent magnets can be built up by considering that they contain magnetic north and south poles, and that the force between two poles is proportional to the product of the pole strengths, and inversely proportional to the square of the distance. It is true that single poles do not seem to exist in nature, but an ordinary magnet can be considered as made up of equal north and south poles in juxtaposition, a combination known as a "dipole."

Coulomb's studies were carried on in the latter half of the eighteenth century. Early in the nineteenth came the discovery of the magnetic effects of continuous currents.   First was Oersted's observation that electric current flowing in a loop of wire exerted magnetic forces on permanent magnets, just as if the loop itself were a magnet. Then came Ampère to formulate these observations mathematically, showing that the magnetic force resulting from a circuit can be broken up into contributions from infinitesimal lengths of wire in the circuit, and that each of these forces falls off as the inverse square of the distance, a law often known by the names of Biot and Savart.   These laws of action at a distance suggested that electromagnetism would develop along the lines suggested by gravitational theory.

Michael Faraday, in the first half of the nineteenth century, was the first who really turned the electromagnetic theory into the lines of field theory.   If a piece of insulator, or dielectric, is put between two charged objects, the force between the objects is diminished.   Faraday was not content to regard this as merely a shielding effect, or a change in the force constant.   He directed attention to the dielectric, and concluded that it became polarized, acquired charges which themselves contributed to the force on other charges.   To describe these effects, he introduced the idea of lines of force, lines pointing in the direction of the force that would be exerted on a charge located at an arbitrary point of space.   He gave a physical meaning to the number of lines per unit area, setting this quantity proportional to the magnitude of the force.   He thought of the lines of force in a very concrete way, as if there were a tension exerted along them, and a pressure at right angles to them, and showed that such a stress system would account for the forces actually exerted on charges.   Faraday's fundamental idea, in other words, was that things of the greatest importance were going on in the apparently empty space between charged bodies, and that electromagnetism could be described by giving the laws of the phenomena in this space, which he called the "field."   His discovery of electromagnetic induction, in which electromotive force is induced in a circuit by the time rate of change of magnetic flux through the circuit, added certainty to his concepts, by pointing out the importance of the magnetic field and its flux.

Faraday was not a mathematician, and his concepts of the field did not immediately appeal to the mathematicians, who were still thinking in terms of inverse-square laws.   His contemporary Gauss furnished the first mathematical formulation of field theory.   Gauss considered lines of force, their flux out of a region, and proved his famous theorem, relating this flux to the total charge within the region.   It remained for Maxwell, however, some thirty years after

Faraday's first discoveries, to find the real mathematical formulation of them. Maxwell accepted wholeheartedly the idea that the electric and magnetic fields were the fundamental entities, and considered the partial differential equations governing those fields. He had a background of experience to work on. In addition to the work of Gauss, there was the formulation of gravitational theory in terms of the gravitational field and potential, which had been worked out at the end of the eighteenth century by Laplace, Poisson, and others. At the time, that formulation seemed more a mathematical device than anything else, but in the hands of Maxwell it furnished an ideal mathematical framework for Faraday's ideas. The electromagnetic field is much more complicated than the gravitational, however, and Maxwell had to go far beyond Laplace, Poisson, and Gauss, introducing among other things the concept of displacement current, which proved to be necessary to reach a mathematically consistent theory. Maxwell's equations have stood the test of time since then, and still furnish the correct formulation of classical electromagnetic theory; it is only the quantum theory which has brought about a fundamental revision of our ideas, during the last few years.

As soon as Maxwell formulated his equations, he was able to draw from them a mathematical result predicting a new phenomenon, which would hardly be suspected from the laws of Coulomb and of Faraday which were his starting points. He was able to show that an electromagnetic disturbance originated by one charged body would not be immediately observed by another, but that instead it would travel out as a wave, with a speed that could be predicted from electrical and magnetic measurements. Furthermore, the velocity so predicted proved to agree, within the small experimental error, with the speed of light. Thus at one blow he accomplished two results of the greatest importance in the history of physics. First, he gave a convincing proof of the superiority of a field theory to action at a distance; secondly, he tied together two great branches of physics, electromagnetism and optics.

To see why action at a distance can hardly explain the propagation of electromagnetic waves, consider as simple a thing as a radio broadcast. In the transmitting antenna, certain charges oscillate back and forth, depending on the signal being transmitted. According to the field theory, these charges produce an electromagnetic wave, which travels out with the speed of light. The wave reaches a receiving antenna an appreciable time later, and sets the charges in that antenna into oscillation, with results that can be detected in the receiver. The forces on the charges in the receiving antenna are not determined at all by the instantaneous positions or velocities of the charges in the transmitting antenna, but by the values that they had at an earlier time. Any reaction back on the transmitter will be delayed by the time taken by the disturbance to reach the receiver, and then to return to the transmitter again, as in an echo. The forces on a particle, in other words, do not depend on the positions of other charges, but on what they did at past times. It is almost impossible to formulate this in terms of action at a distance, but easy to formulate if we regard the electromagnetic field as a real entity, taking energy from the transmitter, and carrying it with a finite velocity to the receiver.

To appreciate the relations between electromagnetism and optics, which Maxwell demonstrated, we have to go back somewhat further with the development of optics. At the time of Newton and Huygens, there were two opposed theories of light, Newton holding a corpuscular theory, in which the light was a stream of infinitesimal particles, being bent as they passed from one medium to another on account of a surface force resulting from different potential energies in the various media; whereas Huygens believed that light was a form of wave motion, and was able to explain reflection and refraction on the basis of the propagation of spherical wavelets, traveling with different velocities in different media. Newton's principal objection to the wave theory was his feeling that it did not explain the way in which obstacles cast sharp shadows. He was thinking by analogy with sound, which was known to be a wave motion, in which sound bends around obstacles. The thing he did not realize was that the wave lengths of light are so small, and that that entirely changes the behavior of shadow formation. It is curious that he did not think of this, for he was familiar with the phenomena of interference and diffraction; he made a theory of them which postulated a periodic disturbance along the direction of wave propagation, which he described as alternate fits of easy reflection and of easy transmission, and by measurement of interference patterns he determined the wave length of this periodic disturbance, in good agreement with modern measurements of the wave length of light. His combination of corpuscles with a periodic disturbance, in fact, showed extraordinary similarity to the present picture resulting from the quantum theory, in which we picture particles, or photons, traveling in accordance with a guiding wave field.

The explanation of interference and diffraction from the wave theory, together with the proper treatment of the casting of shadows, did not actually come until the first years of the nineteenth century, when Young and Fresnel made their discoveries in that field. Fresnel not only explained these phenomena, but also formulated the laws of reflection and refraction, giving laws, which have proved to be correct, for the fraction of the incident light reflected and refracted, as well as for the direction of the reflected and refracted beams. Those laws also explained the phenomenon of double refraction, by which certain crystals such as Iceland spar transmit light in two different rays, the ordinary and the extraordinary rays, traveling at different angles and speeds, and not satisfying the ordinary laws of refraction as found in an isotropic medium. The two rays show properties described as polarization, which proved to result from the fact that light is a transverse, not a longitudinal, vibration, so that two directions of vibration are possible, both at right angles to the direction of propagation. The whole theory of these vibrations shows a close analogy to the transmission of transverse elastic vibrations in an elastic solid, with the one exception that there is no indication of an accompanying longitudinal vibration, such as there would be with an elastic solid, and such as would constitute the only mode of vibration for a fluid.

The physicists of the nineteenth century were much devoted to mechanical models. If light acted like the transverse vibration of an elastic solid, they tried to visualize it as a real solid, and gave it a name, the "ether." It was hard to understand its properties. In the first place, as we have just mentioned, a real elastic solid would transmit longitudinal as well as transverse waves, and no good way of modifying the theory was found that would eliminate the longitudinal waves. In the second place, the solid would have to fill all space, and it was obviously very hard to see how, with a very rigid solid filling space, it was possible for ordinary bodies to move around freely. Much thought was devoted to these questions. Even after Maxwell had shown that light was an electromagnetic disturbance, not a vibration of a solid, there was still much speculation about the nature of the ether. It is really only within the present century that physicists have realized that that speculation is essentially meaningless, that the electric and magnetic fields are the fundamental entities concerned with optics as well as with electromagnetic forces, and that we do not have to endow these fields with mechanical properties foreign to their real nature.

It was this background of an elastic-solid theory of light which

Maxwell encountered when he formulated his electromagnetic theory. That theory at once removed all the difficulties of the previous theories. It yielded only the transverse vibrations, having no solutions of the fundamental equations corresponding to longitudinal waves. Fresnel's equations for refraction and reflection, and his explanations of interference and diffraction, though proposed for an elastic-solid theory, proved to be equally valid in the framework of electromagnetic theory. And the explanation of the casting of shadows, resulting from the small wave length shown to exist by experiments on interference and diffraction, properly answered Newton's objection to the wave theory. Taken together with the remarkable success of the theory in predicting the velocity of light from purely electrical measurements, all doubt about the electromagnetic nature of light almost immediately disappeared, and optics is now treated as a branch of electromagnetism, as we shall treat it in this volume. The argument was placed beyond question a few years later, when Hertz demonstrated the existence of electromagnetic waves of wave lengths of a few centimeters. This was soon followed by the use of much longer electromagnetic waves for radio communication, leading back in the last few years to the use of microwaves, of a few centimeters in length, forming one of the most perfect examples of the application of Maxwell's equations.

The developments we have described comprise most of the classical part of electromagnetic theory, the part that is taken up in the present volume. Two more recent advances have concerned themselves with the quantum theory. First, and more revolutionary, has been the discovery of the photon, and of the fact that light has an aspect that must be explained in a corpuscular way, as well as a wave aspect. This was at first a mathematical deduction by Planck, from the theory of black-body radiation, at the beginning of the present century, and it was followed shortly by Einstein's deduction of the law of photoelectric emission, from Planck's quantum hypothesis. Light of rather short wave length, falling on a suitable metal surface, ejects electrons; these electrons are observed to have an energy depending only on the frequency of the light, being in fact proportional to the frequency, and being independent of the intensity of the light, which regulates merely the number of photoelectrons, not their energy. This suggested strongly that the energy of the light was carried by certain corpuscles, or photons, shown by Einstein to carry a quite finite energy hf, where h is Planck's constant, f the frequency, and that a photon absorbed by an electron of the metal transferred its energy to the electron, so that it was ejected as a photoelectron. Such a hypothesis was in direct conflict with the wave theory, which as we shall see predicts a continuous distribution of energy, and for a number of years this conflict was regarded as a great stumbling block in the development of the quantum theory.

The difficulty still exists, in its way, but it is gradually being worked out in the framework of developing quantum electrodynamics. It is becoming recognized that the corpuscular and the wave point of view simultaneously have their truth, and that the wave is the correct description of the average behavior of the photons, but does not predict the behavior of an individual photon, which in fact seems not to be predictable by any precise theory. The photons move so that on the average they deliver the same energy to any illuminated area that would be predicted by the wave theory, but the energy is actually delivered by the photons, in discrete amounts. At comparatively long wave lengths, or low frequencies, the energy of the photons is so small that a great many are delivered by a source of ordinary energy, and their discreteness is not of importance. At very high frequencies, however, the finite size of the photons is of striking significance, as can be seen in experiments with counters and cloud chambers, in which evidence of a single photon (or of a single electron or ion) can be directly observed.

The less striking, but probably equally important, development of electromagnetic theory in the last few years has been its place in the quantum theory of atomic and molecular structure. Bohr, in 1913, was able to explain the structure of a hydrogen atom, as being made up of a proton and an electron, acting on each other by ordinary electrostatic forces, but governed by quantum mechanics. Developing this theory, Schrödinger in 1926 proposed his wave mechanics, a form of quantum mechanics suggested by the duality between corpuscles and waves in the theory of light. It had been suggested by de Broglie that there was a wave phenomenon associated with the motion of electrons, as there was with the motion of photons, and Schrödinger formulated this wave in terms of a wave equation, somewhat similar to the wave equations of optics. By solving this equation, we find descriptions of great accuracy of the structure of atoms, molecules, and solids and matter of all types. The interesting point is that the forces that enter into Schrödinger's theory are essentially electromagnetic forces as described by classical electromagnetic theory; it is in the mechanics that this theory differs from purely classical theory. Thus we understand the very broad way in which electromagnetic forces underlie a great segment of chemistry, and of the theory of the solid, liquid, and gaseous states.

The applications of Schrödinger's theory to the structure of matter deal almost entirely with particles traveling slowly compared with the velocity of light. The electromagnetic theory that has to be used is then essentially electrostatics, making no use of the finite velocity of propagation of light. When we come to apply the theory to particles traveling with velocities comparable with that of light, we meet essential difficulties, which have not yet been fully solved. We must expect a combination of the theory of radiation, and of the electromagnetic forces concerned in an atom or molecule, which will be consistent and complete, and this does not yet seem to exist. Of course, particles traveling with velocities near that of light must satisfy the relativistic mechanics. Einstein, in the early years of the century, devised his theory of relativity, to explain the dynamics of rapidly moving particles, and the way in which the velocity of light appeared the same in any frame of reference. The modifications that proved to be necessary were only in the dynamics, not in the electromagnetic part of the theory; the forces as predicted by classical electromagnetic theory are relativistically correct, as had been shown earlier by Lorentz. But, in setting up the quantum electrodynamics of rapidly moving particles, it is relativistic rather than Newtonian mechanics which must be used as a start, and it is this combination which is not yet without its difficulties.

Probably part of the trouble with these theories is the fact that new phenomena are appearing as we go to very rapid particles of very high energy, which complicate the whole theory. These are the phenomena of the forces between the particles, neutrons and protons, in the nucleus. There is good evidence that these forces are not electromagnetic, but of another variety, produced not by the electromagnetic field, but by a meson field. This field has been described mathematically by analogy with electromagnetic theory. It similarly has a wave aspect, but also a corpuscular aspect, the corpuscles in this case being mesons, rather than photons. It is thus becoming likely that there exist in nature a number of different levels of forces and particles: the electromagnetic field and the photons, the electrons and protons and neutrons and their associated fields of the de Broglie or Schrödinger type, and the mesons and their fields. And the guiding pattern for the development of all these theories is at present the electromagnetic theory in its classical form. It remains to be seen what final synthesis of these various theories can eventually be made.

We are now ready to proceed with our study of classical electromagnetic theory, which alone we shall take up in this volume. We start in the historical order, treating electrostatics, first from the basis of Coulomb's law, then from the standpoint of field theory, showing as Faraday did how much that helps us in the study of problems involving dielectrics and conductors. Then we take up in a similar way the magnetic field, passing on through the study of electromagnetic induction to Maxwell's equations, and to their application to the electromagnetic theory of light and of other electromagnetic radiation fields.

1. The Force on a Charge.—Electromagnetic theory deals with the forces acting on charges and currents. We find that an electric charge at a given point of space is acted on by two types of force: an electric force, independent of its velocity, and a magnetic force, proportional to its velocity (that is, to the current carried by the charge), and at right angles to its velocity. We find that different charges at the same point of space are acted on by different amounts of force, and we arbitrarily define the strength of the charge as being proportional to the magnitude of force acting on it in a given field. We shall see later how to define the unit of charge, the coulomb. We can define two vectors at every point of space, $\mathbf{E}$ the electric intensity, $\mathbf{B}$ the magnetic induction, such that the force $\mathbf{F}$ on a charge of q coulombs moving with velocity $\mathbf{v}$ is given by the vector equation

$$\mathbf{F} = q(\mathbf{E} + \mathbf{v} \times \mathbf{B}). \tag{1.1}$$

Here $\mathbf{v} \times \mathbf{B}$ is the vector product of $\mathbf{v}$ and $\mathbf{B}$, a vector at right angles to both, whose magnitude is the product of the magnitudes of $\mathbf{v}$ and $\mathbf{B}$ times the sine of the angle between. The reader unfamiliar with this and other aspects of vector notation will find vector methods discussed in Appendix I. To measure $\mathbf{E}$ and $\mathbf{B}$ we need only measure the force on a moving charge at the point in question. The second term is the ordinary motor rule, that the force on a current element is proportional to the current, and the component of magnetic field at right angles to it, and is at right angles to each. In the mks (meter-kilogram-second) system of units, which we shall use, the force will be given by (1.1) in newtons (1 newton = 10^5 dynes), if the charge q is in coulombs, if $\mathbf{E}$ is in volts per meter, $\mathbf{v}$ in meters per second, and $\mathbf{B}$ in webers per square meter (1 weber/sq m = 10^4 gausses). The mks system of units is discussed in Appendix II, in which a discussion is also given of various other commonly used sets of units, and of the form that familiar equations take in these other units. We shall see later how to define the volt and the weber, as well as the coulomb.

As far as charge is concerned, with our knowledge that all matter is composed of electrons, protons, neutrons, and other such elementary particles, all of which prove experimentally to carry charge equal to an integral multiple of the electronic charge e, given by

$$e = 1.60 \times 10^{-19} \text{ coulomb}, \qquad (1.2)$$

we see that the determination of the charge on a body really resolves itself into a counting of the unbalanced elementary charges on it. This fact, that all charges are integral multiples of a fundamental unit, is still one of the unexplained puzzles of fundamental physics. It does not in any way contradict electromagnetic theory, but it is not predicted by it, and until we have a more fundamental theory that explains it, we shall not feel that we really understand electromagnetic phenomena thoroughly. Presumably its explanation will not come until we understand quantum theory more thoroughly than we do at present.

2. The Field of a Distribution of Static Point Charges.—For a number of chapters we shall be dealing with the forces on charges at rest, and shall consider only the electric field **E**, which will be assumed to be independent of time. This is the branch of the subject known as "electrostatics." It is now found that the field **E** can be determined, once the distribution of charge is known. This fact, together with (1.1) and Newton's laws of motion, furnishes a complete system of equations: knowing where the charge is, we find **E**; knowing **E**, we find the force **F**; knowing the force, we find the acceleration of the particles bearing the charge, and hence their motions. Of course, our special case of electrostatics must be that in which the total force, which may include nonelectrostatic as well as electrostatic forces, acting on a body, is zero, so that the body can stay at rest, and we can be dealing with a static problem.

The law giving the field **E** arising from a distribution of charge is very simple. We may consider the charge to be made up of a great many small, or point, charges (which on an atomic scale could be electrons and protons). We may compute the field resulting from each of these point charges. Then the first experimental law is that the total field is the vector sum of the fields arising from the various point charges. Furthermore, the field **E** of a static point charge q proves to be in the direction pointing away from q, and is equal in magnitude to $q/4\pi\epsilon_0 r^2$, where

$$\epsilon_0 = 8.85 \times 10^{-12} \text{ farad/m}, \qquad (2.1)$$

and where r is the distance from the charge, measured in meters.

This may for the present be considered an experimental law, with (2.1) as an experimentally determined coefficient, determined to give the result for **E** in volts per meter, if q is in coulombs. In Appendix II we shall see why it is that in terms of the definition of the volt and the coulomb we have this particular numerical constant. The factor 4π is included in the formula at this point, because it is more convenient to do so for later applications, as we shall see presently. We shall also understand later the units, farads per meter, which we have assigned to ϵ_0 in (2.1).

As a result of our simple law giving the field arising from a point charge, we can find easily the force between two charges q and q', at a distance r apart. The force is clearly in the direction joining them, and is equal in magnitude to

$$F = \frac{qq'}{4\pi\epsilon_0 r^2}. \tag{2.2}$$

It is a repulsion if q and q' have the same sign, an attraction if they have opposite sign. Equation (2.2) is Coulomb's law, in the form that it takes in the mks system; we give corresponding expressions in other systems of units in Appendix II. In Eq. (2.2) we have an illustration of the fact that problems in electrostatics can be handled by the ideas of action at a distance, since Coulomb's law is similar to the law of gravitation.

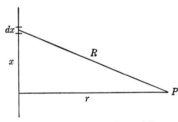

FIG. 1.—Field of a charged line.

It is such a simple matter to find the field of a point charge that we can easily sum such fields vectorially, to find the field of an arbitrary charge distribution, provided that we know where the charges are located. Simple examples of the field of distributions of particles are the fields of a uniformly charged line and plane. Let a line carry a constant charge σ per unit length. The contribution of the charge in length dx to the field at P, in Fig. 1, is along the direction of R; its component along r, which alone integrates to something different from zero, is $(\sigma\,dx/4\pi\epsilon_0 R^2)(r/R) = \sigma r\,dx/4\pi\epsilon_0 R^3$. The resultant field is

$$\frac{\sigma r}{4\pi\epsilon_0} \int_{-\infty}^{\infty} \frac{dx}{(x^2 + r^2)^{3/2}} = \frac{\sigma}{2\pi\epsilon_0 r}.$$

Similarly let a surface carry a constant charge of σ per unit area. The contribution of the charge in the ring of radius between x and $x + dx$ in the plane, shown in Fig. 2, to the component of field along the normal, at P, is $(2\pi\sigma x \, dx/4\pi\epsilon_0 R^2)(r/R) = \sigma rx \, dx/2\epsilon_0 R^3$, and the resultant field is

$$\frac{\sigma r}{2\epsilon_0} \int_0^\infty \frac{x \, dx}{(x^2 + r^2)^{3/2}} = \frac{\sigma}{2\epsilon_0},$$

a constant independent of the distance from the plane. It should be noted, however, that the field is directed away from the plane on each side of the plane (provided σ is positive), so that there is a discon-

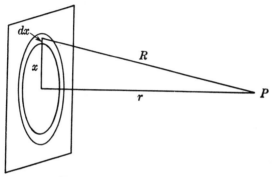

FIG. 2.—Field of a charged plane.

tinuity of σ/ϵ_0 in the normal component of **E** in passing through the plane. This problem of the field resulting from a distribution of charge on a plane illustrates the reason why the factor 4π was included in the statement of Coulomb's law, as in (2.2). The reason is that this factor then drops out in problems such as the field of a plane charge. If we had not included the 4π at first, it would have appeared at this point. Since more problems are similar to that of the field of a plane than to the field of a point charge, it is more convenient to handle the factor 4π as we have done.

The two examples that have just been presented show how simple it is to find fields by direct integration in some cases. We shall now go on to a more involved, but more powerful, method of finding the field of a distribution of charge, using the method of the potential. This method has two advantages over direct integration. First, it is more powerful mathematically; there are many problems to which it will provide solutions, and in which the direct integration proves to be too difficult to carry out. Secondly, there is a whole class of problems in which direct integration is of no use, for we do not know the charge distribution to start with. This class of problems includes those involving conductors, or dielectrics. With such bodies, as we shall see, the presence of certain charges in the neighborhood induces other charges on the conductors or dielectrics, or, as we say, polarizes them. Part of our problem is to find the nature and location of this polarized charge. The method of the potential can handle this problem, whereas direct integration cannot. As a first step in taking up the potential, we consider the work done in moving a point charge from one place to another in the field, and the potential-energy function resulting from this work done.

3. The Potential.—The work that we must do in carrying a unit charge from one point to another in the field of a point charge, balancing the force exerted by the point charge, is independent of the path. To prove this, we note that the work done in going from one point to an adjacent point is the product of the force $-q/4\pi\epsilon_0 r^2$ which we exert to balance the electrostatic force, by the component of displacement dr in the direction of force, or is $-\mathbf{E} \cdot \mathbf{ds}$, where $\mathbf{E}$ is the electric field, $\mathbf{ds}$ is the vector displacement, and where the scalar product $\mathbf{E} \cdot \mathbf{ds}$ equals the product of the magnitudes of $\mathbf{E}$ and $\mathbf{ds}$, and of the cosine of the angle between them, as is discussed in Appendix I. Thus, integrating,

$$\text{Work} = -\int_1^2 \mathbf{E} \cdot \mathbf{ds} = \int_{r_1}^{r_2} -\frac{q}{4\pi\epsilon_0 r^2}\, dr = \frac{q}{4\pi\epsilon_0}\left(\frac{1}{r_2} - \frac{1}{r_1}\right). \quad (3.1)$$

This work may then be written as the difference of a potential energy at the end points: if we write

$$\varphi = \frac{q}{4\pi\epsilon_0 r} \quad (3.2)$$

as the potential function at a distance r from a charge q, the work done moving a unit charge from point 1 to point 2, by (3.1), is

$$\text{Work} = \varphi_2 - \varphi_1.$$

Then, as always with problems involving potentials, we may write the force as the negative gradient of the potential:

$$\mathbf{E} = -\text{grad } \varphi = -\nabla\varphi, \quad (3.3)$$

where the vector operations are discussed in Appendix I. Furthermore, using the principle of vector analysis that the curl of any gradient is zero, we have

$$\text{curl } \mathbf{E} = \nabla \times \mathbf{E} = 0. \quad (3.4)$$

These relations (3.3) and (3.4) hold for the field of a point charge; by addition, they hold for the field of any arbitrary number of point charges.

Use of this principle gives us a simplified way of computing the field of a distribution of charges. We sum the potentials of the charges, so as to get the total potential as a function of position; then we take its gradient, to get the field. This allows us to sum the potential, a scalar, rather than the field, a vector. If charges are distributed continuously in space, instead of at discrete points, we may describe them by a charge density ρ, such that $\rho\, dv$ is the charge located in volume element dv, so that ρ is in coulombs per cubic meter in the mks system. Then we may write the potential of this distribution as

$$\varphi = \frac{1}{4\pi\epsilon_0} \int \frac{\rho\, dv}{r}. \tag{3.5}$$

In this expression, we are finding φ as a function of x, y, and z; the integration is over $dx'\, dy'\, dz' = dv$, the coordinates of the point where the density ρ, which is really a function of x', y', z', is found. r is the distance between these two points, given by

$$r = \sqrt{(x - x')^2 + (y - y')^2 + (z - z')^2}.$$

The potential is expressed in volts. It will automatically satisfy (3.4), from the way in which it is built up. From φ as given in (3.5), we then find $\mathbf{E}$ by (3.3).

Surfaces $\varphi = $ constant are called "equipotential surfaces." The field $\mathbf{E}$ is normal to the equipotential surfaces, as we can see either from the principle of vector analysis that the gradient of a scalar is normal to the surfaces on which the scalar is constant, or from the elementary fact that, since the work done on a charge moving from one point to another of the same equipotential surface must be zero, any displacement on that surface must be at right angles to the force vector. We may draw lines, called "lines of force," everywhere tangent to $\mathbf{E}$; they then cut the equipotentials at right angles. These concepts, of the sort stressed particularly by Faraday, are particularly useful in electrostatic problems involving conductors. In a conductor, charges are free to move about. Thus, if conductors are present in a field, we cannot predict straightforwardly where the charge is to be found, and hence cannot find φ by (3.5), since we do not know ρ. We do know, however, that a conductor satisfies Ohm's law, which states that the current in a conductor is proportional to the electric field in it. If our problem is a static one, in which all charge is at rest and there is no current, the field must then be zero everywhere within the conductor. The potential must then be constant, by (3.3). Thus a conductor forms an equipotential volume, its surface being an equipotential surface, so that all lines of force must cut it at right angles.

4. Electric Images.—Suppose we are given the problem of finding the potential, and hence the field, of a certain set of charges, q_1, q_2, . . . , at specified points, in the presence of certain conductors that are maintained at definite potentials. The potential must then

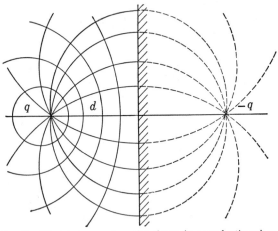

Fig. 3.—Electric image of a point charge in a conducting plane.

satisfy the following conditions: it must be equal to the required constant value on the surface of each conductor, and it must reduce to a value like (3.2) at each of the charges. These conditions can be shown to determine the potential uniquely. Sometimes by a simple device we can set up a potential, or the corresponding field, satisfying these conditions; we can then be sure we have solved our problem. Thus consider a single charge q at a distance d from a grounded conducting plane. There is a problem, shown in Fig. 3, which has a potential that reduces to the proper value where the charge is, and which gives a potential zero on the surface, by symmetry: it is the problem of two equal and opposite charges $\pm q$, at a distance $2d$ apart, the mid-plane taking the place of the surface. The correct lines of force and equipotentials are then as shown in Fig. 3, the dotted extensions being really nonexistent. The fictitious charge $-q$ behind the metallic surface is called an "electric image," from the analogy of an optical image, and this method is called the "method of images."

The method of images can also be used for the problem of a point charge and a conducting sphere. This depends on a geometrical theorem. In Fig. 4, let $oa/oc = oc/ob$. The triangles oac and obc are

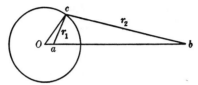

Fig. 4.—Image of a point charge in a conducting sphere.

similar, for they have the angle aoc in common, and the sides including the angle are proportional. Thus

$$\frac{oa}{oc} = \frac{oc}{ob} = \frac{r_1}{r_2}. \tag{4.1}$$

If we then place a charge q at b, and a charge $-(oc/ob)q$ at a, the potential at any point of the circle will be

$$\frac{q}{r_2} - \left(\frac{oc}{ob}\right)\frac{q}{r_1} = \frac{q}{r_2} - \frac{q}{r_2} = 0,$$

using (4.1). Thus the circle, or the sphere formed by rotating it around the axis ob, is an equipotential of zero potential. Hence the two charges, q at b and its image $-(oc/ob)q$ at a, give a field that is a solution of the problem. A somewhat more general problem can be solved by adding to the field of these two charges the field of an arbitrary point charge at the center o of the circle. This charge will give a potential that is constant over the surface of the sphere, so that the field of all three charges is consistent with the assumption that the sphere is a conductor. By adjusting the amount of charge at the center o, we can solve the problem of the potential of a point charge q at b, and a conducting sphere carrying any desired amount of charge.

Problems

1. An electron of charge $-e$ ($e = 1.60 \times 10^{-19}$ coulomb) and mass m ($m = 9.1 \times 10^{-31}$ kg) moves in a magnetic field B (in gausses) in a plane at right angles to B. Show that it moves in a circular path of radius ρ, and find the value of the product $B\rho$ in terms of the velocity of the electron.

2. An electron moves in a uniform electric field directed along the x axis, and a uniform magnetic field along the y axis. It starts from rest at the origin. Show that it moves in a cycloidal path, and find the drift velocity, or the velocity of the center of the rolling circle along the z axis. If the electric field is 10^4 volts/cm, and the magnetic field is 10^4 gausses, find the value of the drift velocity.

3. Find the angular velocity of rotation of an ion in a cyclotron. If the magnetic field is 15,000 gausses, what should be the wave length and frequency of the oscillator used to run the cyclotron, if it is accelerating protons?

4. When the velocity of a particle approaches that of light, a force at right angles to the velocity produces an acceleration transverse to the velocity given by the equation $F = m_t a$, where m_t is the transverse mass, equal to

$$\frac{m_0}{\sqrt{1 - v^2/c^2}},$$

if m_0 is the rest mass, v the velocity, c the velocity of light. Discuss the cyclotron in the relativistic range, showing that the resonant frequency must change as the ions are speeded up. If protons are to be speeded up to 500 million electron volts, find the ratio by which the oscillator frequency must change as they accelerate from rest to their maximum speed.

5. A charge e is located a distance d from an infinite conducting plane. Find how much work is required to remove the charge to infinite distance, against the attraction of its electric image.

6. Find the potential at distance r from an infinitely long straight wire carrying a charge σ per unit length.

7. Inside a hollow infinitely long grounded conducting cylinder of radius R is placed a thin charged wire, carrying a charge σ per unit length. The wire is placed parallel to the axis of the cylinder, and distant by an amount d from the axis. Find the potential at points within the cylinder. [*Hint:* Use the construction of Fig. 4 and of Eq. (4.1).]

8. Find the attraction between a charge q, and a grounded conducting sphere, as a function of the distance of q from the center of the sphere.

9. Find the attraction between a charge q, and an uncharged, insulated, conducting sphere, as a function of the distance of q from the center of the sphere. (We shall prove later that, if the sphere is uncharged, the total fictitious charge located at points o and a in Fig. 4 must be zero.)

10. Two parallel wires carry equal and opposite charges $\pm\sigma$ per unit length. Consider the intersections of the equipotentials with a plane normal to the wires, and the lines of force, which will lie in that plane. Show that the lines of force and equipotentials form two families of circles, orthogonal to each other.

11. A charge q is placed inside a hollow metal sphere of radius a at a distance r from the center. Show that the force acting on the charge q is given by

$$F = \frac{q^2 a r}{4\pi\epsilon_0 (a^2 - r^2)^2}.$$

12. Find the force per unit length with which two long parallel metal cylinders, each of diameter a and separated by a distance d between centers, attract each other, if they carry a charge per unit length $\pm\sigma$, respectively.

CHAPTER II

ELECTROSTATICS

The simple concept of lines of force and of equipotential surfaces, which we introduced in the preceding chapter, becomes of real value when it forms the basis of an analytical treatment founded on differential equations. This treatment, worked out originally by Gauss, Poisson, Laplace, and others, can be used either for electrostatics, to which we apply it, or for gravitational forces, which consist of inverse-square forces, as electrostatic forces do. It was actually for the gravitational case that much of the mathematical work of the present chapter was originally developed. We shall find, however, that it forms a very natural mathematical framework for electrostatics, and that it leads us into certain equations that will later underlie the general case of electromagnetism, not merely the static problem. We develop the equations in the present chapter, and solve some electrostatic problems by means of the solution of partial differential equations in the next chapter.

1. Gauss's Theorem.—Suppose we set up a closed surface S, enclosing a volume V within which are certain charges q_1, q_2, ... We now form a surface integral over S, as follows: We take an element da of the area of the surface. If $\mathbf{n}$ is unit vector along the outer normal, and if $\mathbf{E}$ is the value of the electric intensity at da, the surface integral is $\mathbf{E} \cdot \mathbf{n}\, da$, integrated over the surface S. Gauss's theorem is then the following:

$$\int \mathbf{E} \cdot \mathbf{n}\, da = \sum_i \frac{q_i}{\epsilon_0}. \tag{1.1}$$

That is, the surface integral of the normal component of $\mathbf{E}$ over a surface S equals $1/\epsilon_0$ times the total charge included within the surface. Let us prove this important theorem. First we prove it for a single point charge; then by addition the theorem obviously holds for any collection of point charges. For a single charge, consider an element da of surface, as shown in Fig. 5. The contribution of this element to $\mathbf{E} \cdot \mathbf{n}\, da$ is $|E|\, da \cos (\mathbf{E}, \mathbf{n})$. But $da \cos (\mathbf{E}, \mathbf{n})$ is the projection of da on the plane normal to $\mathbf{E}$, or is $r^2\, d\omega$, where $d\omega$ is the solid angle of the cone subtended by da. Also $|E| = q/4\pi\epsilon_0 r^2$. Thus the contribution of the element to the surface integral is

$$\frac{q}{4\pi\epsilon_0 r^2}\, r^2\, d\omega \;=\; \frac{q}{4\pi\epsilon_0}\, d\omega.$$

Integrating over S involves integrating $d\omega$ over the complete solid angle 4π, resulting in q/ϵ_0, in agreement with (1.1). Thus, summing

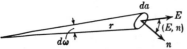

over an arbitrary number of point charges, Gauss's theorem is proved.

Fig. 5.—Construction for Gauss's theorem.

Examples of the application of Gauss's theorem are obvious.

First consider a positive point charge, and a sphere of radius r surrounding it. By symmetry, the field points out along the radius, and depends only on r. Thus Gauss's theorem states that $|E|$ times the area, $4\pi r^2$, equals q/ϵ_0, or $|E| = q/4\pi\epsilon_0 r^2$. This is trivial, but we can equally well use Gauss's theorem to find the field of a uniformly charged line and plane, as in Chap. I, Sec. 2. For a line carrying a charge σ per unit length, let the surface S be the surface of a circular cylinder of unit length, radius r, with the line as the axis. By sym-

metry, the field points radially outward, and depends only on r. Thus the field is parallel to the flat ends of the cylinder, which do not contribute to the surface integral; while over the curved face the integral is $|E|$ times the area $2\pi r$. Since this equals σ/ϵ_0, we have $|E| = \sigma/2\pi\epsilon_0 r$. For the charged plane, take S as the surface of a box, as shown in Fig. 6,

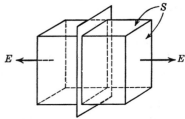

Fig. 6.—Gauss's theorem for field of charged plane.

enclosing unit area of the surface. By symmetry, **E** will be normal to the two faces parallel to the plane. Hence only these faces will contribute to the integral. Each has unit area; thus $2|E| = \sigma/\epsilon_0$, $|E| = \sigma/2\epsilon_0$. Thus we verify the results of Chap. I, Sec. 2, by very simple methods.

As a final example, consider the surface of a conductor carrying a surface charge σ per unit area. We set up a surface as in Fig. 6, but now to the left of the surface, or inside the conductor, E is zero. Our volume still encloses charge σ, but now the surface integral has a contribution only from the right-hand face. Thus $|E| = \sigma/\epsilon_0$ on the right. We note that this case can be found from the earlier one by superposing a constant field $\sigma/2\epsilon_0$ pointing to the right on our earlier solution. This field cancels the field originally present to the left of the plane, and doubles the field originally present on the right. Such a constant field makes no contribution to the surface integral concerned in Gauss's theorem, and in fact any constant field, or more generally the field of any charge distribution outside the surface, may be added to our original solution of the problem, without contradicting Gauss's theorem. It is thus clear that Gauss's theorem by itself is not enough to determine the field completely. In our first case of the charged plane we used the additional fact of symmetry, and in the second case of the charged surface of the conductor we used the fact that the field must be zero within the conductor. In any actual case there would be information sufficient to determine the field definitely. We note an important result common to all these possible solutions for the charged plane, however: the normal component of **E** undergoes a discontinuity of σ/ϵ_0 in every case in going through a surface charge σ. Thus in the case of the charged plane the component to the right of the field changes from $-\sigma/2\epsilon_0$ to $\sigma/2\epsilon_0$, and in the case of the conducting surface it changes from zero to σ/ϵ_0. It is clear from Gauss's theorem that this relation is quite general. It has a valuable application, which will be brought out in the problems: if we know the potential and field at every point of space, including the neighborhood of conductors, we can calculate the surface-charge density on the surface of the conductors, since we know the normal component of field. We thus see, as we pointed out in the preceding chapter, that a solution for a potential that satisfies the correct boundary conditions on the surface of all conductors allows us to find the distribution of charge, though otherwise we should not know how the charge was distributed.

2. Capacity of Condensers.—A condenser consists of two conductors carrying equal and opposite charges; its capacity C is defined as the charge on one of the conductors, divided by the difference of potential between them. In a number of simple cases we can easily get the capacity by use of Gauss's theorem. Thus consider two parallel plates of area A, distance of separation d. The field **E** will be normal to the surfaces (if we neglect edge effects). If there is surface charge σ on one plate, $-\sigma$ on the other, there will be a field $E = \sigma/\epsilon_0$ between; and the difference of potential, or voltage V, between them, will be $Ed = \sigma d/\epsilon_0$. The charge Q on the plates will be σA. Thus the capacity will be

$$C = \frac{\sigma A}{\sigma d/\epsilon_0} = \frac{\epsilon_0 A}{d}.$$

From this we see the physical meaning of ϵ_0: it is the capacity of a condenser whose area is 1 sq m, with two plates 1 m apart.

Next consider a cylindrical condenser, consisting of two concentric cylinders of radii r_1, r_2 $(r_2 > r_1)$, of length L. Let the inner conductor carry charge $-Q$, the outer one Q. The field between the two, by symmetry, must be radially inward, and hence by Gauss's theorem must be $-(Q/L)/2\pi\epsilon_0 r$, just as if the charge were concentrated along the axis. The potential at distance r is then the quantity whose negative gradient is the field, or is $[(Q/L)/2\pi\epsilon_0] \ln r$. Thus the difference of potential is

$$\left[\frac{Q/L}{2\pi\epsilon_0}\right] (\ln r_2 - \ln r_1) = \left[\frac{Q/L}{2\pi\epsilon_0}\right] \ln \frac{r_2}{r_1}$$

and the capacity is

$$C = \frac{2\pi\epsilon_0 L}{\ln r_2/r_1}.$$

For a spherical condenser, consisting of two concentric spheres of radii r_1 and r_2 $(r_2 > r_1)$, carrying charges $-Q$ and Q, respectively, the field must be radially inward, and hence by Gauss's theorem must be $-Q/4\pi\epsilon_0 r^2$, as if the charge were concentrated at the center. The potential at distance r is then $-Q/4\pi\epsilon_0 r$, and the difference of potential between conductors is $(Q/4\pi\epsilon_0)[(1/r_1) - (1/r_2)]$, so that the capacity is

$$C = \frac{4\pi\epsilon_0}{(1/r_1) - (1/r_2)}.$$

It is interesting to note that, as $r_2 \to \infty$, the capacity stays finite; this value is often referred to as the capacity of a sphere. Thus a

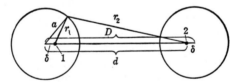

Fig. 7.—Capacity of parallel cylinder.

sphere 1 m in radius has a capacity $4\pi\epsilon_0$ with respect to an infinitely distant sphere.

As a final example of capacity we take a somewhat more complicated case, two parallel cylinders, each of radius a, length L, at distance D apart between their centers, where we assume the length to be so great that we can neglect end effects. We shall first show that equal and opposite charges $\pm Q$ distributed uniformly over the two lines 1 and 2 in Fig. 7, where $\delta/a = a/d$, will result in the two cylindrical conductors being equipotentials. To prove this, we use the geometrical theorem of Eq. (4.1), Chap. I, though for quite a different purpose. From it we see that $\delta/a = a/d = r_1/r_2$. A line charge $+Q$ at 1, and $-Q$ at 2, will have a potential equal to

$$\frac{Q/L}{2\pi\epsilon_0}(\ln r_1 - \ln r_2) = \frac{Q/L}{2\pi\epsilon_0}\ln\frac{r_1}{r_2}$$
$$= \frac{Q/L}{2\pi\epsilon_0}\ln\frac{a}{d}$$

at every point of the left-hand cylinder, and the negative of this at every point of the right-hand cylinder. Thus the cylinders are equipotentials, and can be replaced by conductors. By Gauss's theorem, the total charge on each cylinder is $\pm Q$; for the surface integral of the normal component of **E** over the surface of the cylinder must be $1/\epsilon_0$ times the total charge inside, which is $\pm Q$ for the line charges, and hence for the cylinders that produce equal fields at external points. Thus the difference of potential between the cylinders is $[(Q/L)/\pi\epsilon_0]\ln d/a$, so that the capacity is

$$C = \frac{\pi\epsilon_0 L}{\ln d/a}.$$

This can be put in terms of D, the distance between centers, rather than d. We have

$$\frac{D}{a} = \frac{d}{a} + \frac{\delta}{a} = \frac{d}{a} + \frac{a}{d} = e^{\pi\epsilon_0 L/C} + e^{-\pi\epsilon_0 L/C} = 2\cosh\frac{\pi\epsilon_0 L}{C},$$

or

$$C = \frac{\pi\epsilon_0 L}{\cosh^{-1} D/2a}. \tag{2.1}$$

3. Poisson's Equation and Laplace's Equation.—Suppose that within a volume V we have, not a discrete set of point charges, but a continuous distribution of volume charge, such as we should have for instance from an electronic space charge, if our scale of measurement is large enough compared with the interelectronic distance. We can then define a charge density ρ as the charge per unit volume (in coulombs per cubic meter in the mks system). Then Gauss's theorem is replaced by

$$\int_S \mathbf{E} \cdot \mathbf{n}\, da = \frac{1}{\epsilon_0}\int_V \rho\, dv, \tag{3.1}$$

where the volume integral is over the volume V enclosed by the surface

S. This must hold for any arbitrary surface and volume. Now we may apply the divergence theorem of vector analysis, discussed in Appendix I. This states that, if $\mathbf{F}$ is any vector function of position, and div $\mathbf{F}$, or $\nabla \cdot \mathbf{F}$, is its divergence, then

$$\int_S \mathbf{F} \cdot \mathbf{n} \, da = \int_V \text{div } \mathbf{F} \, dv. \tag{3.2}$$

Applying to (3.1), we have

$$\frac{1}{\epsilon_0} \int_V \rho \, dv = \int_V \text{div } \mathbf{E} \, dv,$$

where the equation holds for any arbitrary volume V. This cannot be the case unless the integrands are equal; that is, unless

$$\text{div } \mathbf{E} = \frac{\rho}{\epsilon_0}. \tag{3.3}$$

This is one of the fundamental equations of electromagnetic theory. We now combine with Eq. (3.3) of Chap. I, $\mathbf{E} = - \text{grad } \varphi$, where φ is the potential. We then have

$$\nabla^2 \varphi = - \frac{\rho}{\epsilon_0}. \tag{3.4}$$

This is Poisson's equation. In a region where there is no volume charge, so that $\rho = 0$, Poisson's equation reduces to

$$\nabla^2 \varphi = 0, \tag{3.5}$$

which is Laplace's equation, and is one of the most important equations in mathematical physics. It is as a result of its appearance in this equation that the operator ∇^2 is called the "Laplacian."

As an illustration of the variety of problems in which Laplace's equation is encountered, we may mention static elasticity; elastic vibrations are governed by a wave equation, and when the time variation in the wave equation is set equal to zero, Laplace's equation results. Another illustration is the steady-state flow of heat, where the temperature satisfies Laplace's equation. Diffusion, and the steady flow of electricity, are two other examples. Many more complicated equations, such as the wave equations of wave mechanics, contain the Laplacian operator, and are solved by analogy with Laplace's equation, so that a knowledge of the solutions of Laplace's equation proves to underlie a great deal of mathematical physics.

4. Green's Theorem, and the Solution of Poisson's Equation in an Unbounded Region.—A partial differential equation, like Poisson's or Laplace's equation, has a great variety of solutions. In the first place, Poisson's equation, being a linear inhomogeneous differential equation (that is, containing the Laplacian, involving first powers of derivatives of φ, and containing the term $-\rho/\epsilon_0$ involving terms independent of φ), has a general solution that can be formed as follows: we find a particular solution, or particular integral, of the inhomogeneous equation. We also find a general solution of the related homogeneous equation, formed by setting the term independent of φ equal to zero [that is, in this case, of Laplace's equation (3.5)]. This homogeneous equation is called the "auxiliary equation," and a general solution of it is called the "complementary function." Then the sum of the particular integral, and of the complementary function, is in the first place a solution of the inhomogeneous equation, as we can see at once by substituting it in the equation. Furthermore, if the complementary function has sufficient arbitrariness in it, we can use it to satisfy arbitrary boundary conditions. Thus the sum of complementary function and particular integral is a complete solution of the inhomogeneous equation. In the next chapter we shall take up some methods for solving Laplace's equation, and satisfying arbitrary boundary conditions with it. We shall find that, to do this, we require an infinite number of arbitrary constants, or an arbitrary function, since we are dealing with a partial differential equation. In the present chapter we shall demonstrate a simple method, called "Green's method," for finding a particular integral of (3.4). By combining the two, we shall then be in position to get a complete solution of Poisson's equation.

The first step in deriving Green's solution of Poisson's equation is to prove Green's theorem, an important theorem in vector analysis, discussed in Appendix I. We prove it from the divergence theorem, (3.2), by setting $\mathbf{F} = \varphi \operatorname{grad} \psi$, where φ and ψ are scalars. Then (3.2) becomes

$$\int_S \varphi \operatorname{grad} \psi \cdot \mathbf{n} \, da = \int_V (\varphi \nabla^2 \psi + \operatorname{grad} \varphi \cdot \operatorname{grad} \psi) \, dv. \quad (4.1)$$

This is one form of Green's theorem. To get the more familiar form, we next write the same expression with φ and ψ interchanged, and subtract, obtaining

$$\int_S (\varphi \operatorname{grad} \psi \cdot \mathbf{n} - \psi \operatorname{grad} \varphi \cdot \mathbf{n}) \, da = \int_V (\varphi \nabla^2 \psi - \psi \nabla^2 \varphi) \, dv. \quad (4.2)$$

We can now use this theorem to obtain Green's solution of Poisson's equation. Let us apply the theorem to the whole of space outside a very small sphere of radius R surrounding a point P. Let $\psi = 1/r$, where r is the distance from P to the point x, y, z. Then $\nabla^2\psi = 0$ everywhere outside the sphere of radius R; for ψ is the potential of a charge located at P, and hence satisfies Laplace's equation everywhere except where the charge is located, or at P. The right side of (4.2) is then

$$- \int \frac{\nabla^2\varphi}{r}\, dv,$$

integrated over all space except the small sphere. For the left side, grad $\psi \cdot n = -d(1/r)/dr = 1/r^2$. Thus the left side is

$$\int \frac{\varphi}{r^2}\, da + \int \frac{1}{r}\frac{\partial\varphi}{\partial r}\, da.$$

On the surface of the sphere, $r = R$, $\int(\varphi/r^2)\, da$ is then

$$\left(\frac{1}{R^2}\right) \bar{\varphi} \int da = 4\pi\bar{\varphi},$$

where $\bar{\varphi}$ is the average value of φ over the sphere, or is approximately the value of φ at P. The second term is $(1/R)(\overline{\partial\varphi/\partial r})(4\pi R^2)$, which goes to zero as R approaches zero. Thus in the limit as R goes to zero, we have

$$4\pi\varphi = - \int \frac{\nabla^2\varphi}{r}\, dv. \tag{4.3}$$

This is a mathematical theorem holding for any function φ, where the volume integral on the right side is extended over all space except an infinitesimal sphere surrounding the point P, where φ is the value of φ at P, and where r is the distance from P to x, y, z. Now let us combine this with Poisson's equation (3.4). Then, for the potential φ of the charge ρ, we have

$$\varphi = \frac{1}{4\pi\epsilon_0} \int \frac{\rho}{r}\, dv. \tag{4.4}$$

This is Green's solution of Poisson's equation. It is identical with Eq. (3.5) of Chap. I, which we have already obtained; it is simply a more elegant method of deriving that result. We shall find this more elegant method to be useful, in the next chapter, in finding the potential in a bounded region of space, rather than an unbounded region as we have here.

5. Direct Solution of Poisson's Equation.—Green's method forms one way of solving Poisson's equation. However, sometimes ρ has a sufficiently simple form so that we can solve it directly, regarding it as a problem in differential equations. Thus, for instance, suppose ρ is a function of one variable only, say x, independent of y and z. Then the potential may also be chosen to be independent of y and z. In this case Poisson's equation becomes an ordinary differential equation:

$$\frac{d^2\varphi}{dx^2} = -\frac{\rho(x)}{\epsilon_0}.$$

This can be solved as an ordinary differential equation for φ as a function of x, merely by integrating twice with respect to x. As a very simple case, let ρ be a constant. Then we have simply

$$\varphi = \varphi_0 - E_0 x - \frac{1}{2}\frac{\rho}{\epsilon_0}x^2. \tag{5.1}$$

Such solutions of Poisson's equation are useful in space-charge problems. In studying space-charge limited emission from cathodes of various shapes, we use Poisson's equation, expressed in variables appropriate to the cathode geometry, to find the potential from the charge distribution. This is combined with the dynamical equations of motion of the electrons, Newton's law giving the acceleration in terms of the force derived from the potential, and in this way we get a complete set of equations to determine both charge density and potential.

We must remember, as has been mentioned earlier, that the solution, as found for instance in (5.1), is by no means a general solution of Poisson's equation, even though it contains the two arbitrary constants, φ_0 and E_0. For it is clear that we can add to it any solution of Laplace's equation, acting as the complementary function. It is true that our function $\varphi_0 - E_0 x$ is a solution of Laplace's equation, but it is far from a general solution, since it is a function of x only. We shall examine general solutions of Laplace's equation in the next chapter, and shall show that a great variety of such solutions exists. We shall show that, by using a general solution of Laplace's equation, we can satisfy general boundary conditions: we can find a solution such that φ reduces to prescribed values over the boundary of the region in which we can satisfy Poisson's or Laplace's equation.

Problems

1. Given a spherical distribution of charge, in which the density is a function of r. Prove that the field at any point is what would be obtained by imagining a sphere drawn through the point, with its center at the origin, all the charge within the sphere concentrated at the center, and all the charge outside removed. Apply this result to the gravitational case, showing that the earth acts on bodies at its surface as if its mass were concentrated at the center.

2. Given a sphere filled with charge of constant density. Prove that, at points within the sphere, the field is directly proportional to the distance from the center.

3. Find the surface density induced by a charge on a plane conductor. Show by direct integration that the total induced surface charge equals the inducing charge in magnitude.

4. For a certain spherical distribution of charge, the potential is given by $\frac{qe^{-ar}}{4\pi\epsilon_0 r}$, where q, a are constants. Find a distribution of charge that will produce this potential, finding charge density as function of r, and the charge contained between r and $r + dr$. Consider whether a point charge at the origin is also required to produce the potential. The resulting charge distribution represents roughly the charge distribution within an atom.

5. There are certain charges and conductors in an electrostatic field, whose potential is φ. Show that the surface density of charge on the surface of a conductor is $-\epsilon_0 \frac{\partial \varphi}{\partial n}$, where n is the normal pointing out of the conductor.

6. Prove that the potential cannot have a minimum in an uncharged region of space. Prove therefore that a point charge cannot be in stable equilibrium under the action of electrostatic forces in an uncharged region.

7. A certain vacuum tube contains a cylindrical cathode of radius r_1. Surrounding it is a space-charge sheath, of constant density $-\rho$ (where ρ is positive), extending to a larger radius r_2. The anode is a cylinder of still larger radius r_3. If the cathode is at potential φ_1, the anode at potential φ_3, find the potential as a function of r, both inside and outside the space charge.

8. Discuss the charge distribution giving rise to equipotentials given by the real part of $-\frac{q}{2\pi\epsilon_0} \ln [\sin \pi(x + jy)] = \text{constant}$. What is the form of these equipotentials for large values of y? Sketch the lines of force and the equipotentials.

CHAPTER III

SOLUTIONS OF LAPLACE'S EQUATION

A partial differential equation, such as Laplace's equation, has a great variety of solutions. The type of solution to be used depends largely on the type of boundary condition that must be satisfied. One method of solution is an extension of Green's method; we meet it later in this chapter. Another method is called "separation of variables." This involves finding solutions that are the product of a separate function of each of the three coordinates. It is a common method for solution of similar equations, and is in fact the most powerful method available for the purpose. Different solutions can be found, depending on whether we work in rectangular, polar, or other coordinate systems. We start with rectangular coordinates, for a simple illustration of the method.

1. Solution of Laplace's Equation in Rectangular Coordinates by Separation of Variables.—Working in rectangular coordinates x, y, z, let us try to find a solution of $\nabla^2\varphi = 0$ in which φ is a product of a function X of x, a function Y of y, and a function Z of z (this is certainly not the most general solution, but we shall find that solutions of this type exist). Then Laplace's equation becomes

$$YZ \frac{d^2X}{dx^2} + ZX \frac{d^2Y}{dy^2} + XY \frac{d^2Z}{dz^2} = 0.$$

We divide by XYZ, obtaining

$$\frac{1}{X} \frac{d^2X}{dx^2} + \frac{1}{Y} \frac{d^2Y}{dy^2} + \frac{1}{Z} \frac{d^2Z}{dz^2} = 0.$$

In this equation, the first term is a function of x only, the second a function of y only, the third a function of z only. It is clearly impossible that the sum of these should be zero independent of x, y, z, unless each term separately is a constant, and the constants add to zero. That is, we must have

$$\frac{1}{X} \frac{d^2X}{dx^2} = a^2, \quad \frac{1}{Y} \frac{d^2Y}{dy^2} = b^2, \quad \frac{1}{Z} \frac{d^2Z}{dz^2} = c^2, \quad a^2 + b^2 + c^2 = 0. \quad (1.1)$$

From these we then have three ordinary differential equations,

$$\frac{d^2X}{dx^2} = a^2X, \qquad \frac{d^2Y}{dy^2} = b^2Y, \qquad \frac{d^2Z}{dz^2} = c^2Z,$$

whose solutions are

$$X = A_1e^{ax} + A_2e^{-ax},$$
$$Y = B_1e^{by} + B_2e^{-by},$$
$$Z = C_1e^{cz} + C_2e^{-cz},$$

where as a result of (1.1), some of the constants a, b, c must be real and some imaginary. Thus some of the functions X, Y, Z vary exponentially with the arguments, the others sinusoidally. The product XYZ is now a particular solution of Laplace's equation.

Any linear combination of such particular solutions is also a solution. Thus as a general solution we have

$$\varphi = \Sigma(A_1e^{ax} + A_2e^{-ax})(B_1e^{by} + B_2e^{-by})(C_1e^{cz} + C_2e^{-cz})$$

where the summation is over an infinite number of terms, each with separate constants A_1, A_2, B_1, B_2, C_1, C_2, a, b, c, subject only to the relation $a^2 + b^2 + c^2 = 0$. Thus the general solution has an infinite number of arbitrary constants. This is characteristic of partial differential equations. The constants are to be chosen so as to fit the boundary conditions. We shall use rectangular coordinates if the boundary conditions are imposed over the surface of a rectangular region. For instance, suppose we have two planes, $x = 0$ and $x = d$, on which the potentials are given as functions of y, independent of z; we might have a composite electrode on each of these planes, different strips being maintained at different potentials by batteries. Suppose we wish to find the potential in the region between the planes. Clearly we can let φ be independent of z, so that $c = 0$, and $a^2 = -b^2$. Then we may write

$$\varphi = \Sigma(A_1e^{ax} + A_2e^{-ax})(B_1e^{jay} + B_2e^{-jay}).$$

where $j = \sqrt{-1}$. When $x = 0$, this becomes

$$\Sigma(A_1 + A_2)(B_1e^{jay} + B_2e^{-jay}),$$

and when $x = d$ it is

$$\Sigma(A_1e^{ad} + A_2e^{-ad})(B_1e^{jay} + B_2e^{-jay}).$$

Now a sum of an infinite number of terms of the form ΣDe^{jay} is one way of writing a Fourier series, which we discuss in Appendix III. Thus at $x = 0$ the potential must be given by a Fourier series with coefficients $(A_1 + A_2)B_1$ for the term in e^{iay}, and at $x = d$ it is given by another Fourier series with coefficients $(A_1 e^{ad} + A_2 e^{-ad})B_1$. By the theory of Fourier series we can find these coefficients, and hence the quantities $A_1 B_1$, $A_2 B_1$ (and similarly $A_1 B_2$, $A_2 B_2$). Thus we can get the infinite set of arbitrary constants, from the boundary conditions.

The method we have used above, for satisfying boundary conditions in Laplace's equation, can be used also in discussing Poisson's equation. Thus suppose we have the same problem as above, two parallel planes on which the potentials are to be prescribed functions of y, but that between the planes we have certain charge distributions that are assumed given. To keep the simplicity of our problem resulting from the independence of z, these charge distributions as well should be independent of z. We now know, from the two preceding chapters, how to find a particular solution of Poisson's equation for the potential resulting from these charge distributions. In particular, for this problem, the charge can be considered as made up of uniform charges along lines parallel to the z axis, and we have already seen that such a charged line has a logarithmic potential, so that all we need to do is to sum such logarithmic potentials arising from all linear charges. The resulting potential, however, will not have the correct values along the two planes. We then set up the complementary function, a solution of Laplace's equation, built up as in the solution we have worked out in this section, and reducing to boundary values on the surfaces which, added to the potential already found from our particular solution of Poisson's equation, will give the desired values. This is a problem of the type we have already considered, so that we see that the determination of the complementary function for solving Poisson's equation is a problem exactly analogous to solving Laplace's equation directly.

2. Laplace's Equation in Spherical Coordinates.—The method of separation of variables, which we have applied in rectangular coordinates in the preceding section, can be applied as well in a number of other coordinate systems, notably in cylindrical and spherical coordinates, and as a less familiar case in ellipsoidal coordinates. The case of spherical coordinates is the more important in practice, for it is used for the field of point charges, dipoles, and other charge distributions concentrated near a point. In spherical polar coordinates r, θ, φ, Laplace's equation for the potential ψ (we use this symbol, so as not to confuse it with the coordinate φ) is

$$\frac{1}{r^2}\frac{\partial}{\partial r}\left(r^2 \frac{\partial \psi}{\partial r}\right) + \frac{1}{r^2 \sin \theta}\frac{\partial}{\partial \theta}\left(\sin \theta \, \frac{\partial \psi}{\partial \theta}\right) + \frac{1}{r^2 \sin^2 \theta}\frac{\partial^2 \psi}{\partial \varphi^2} = 0,$$

as we see in Appendix IV, in which we discuss vector operations in curvilinear coordinates. We assume that ψ is a product of three functions, one a function of r, one a function of θ, one a function of φ. That is, we assume $\psi = R(r)\Theta(\theta)\Phi(\varphi)$. Inserting this function, and dividing by the product $R\Theta\Phi$, we have

$$\frac{1}{R}\frac{1}{r^2}\frac{d}{dr}\left(r^2\frac{dR}{dr}\right) + \frac{1}{\Theta}\frac{1}{r^2\sin\theta}\frac{d}{d\theta}\left(\sin\theta\frac{d\Theta}{d\theta}\right) + \frac{1}{\Phi}\frac{1}{r^2\sin^2\theta}\frac{d^2\Phi}{d\varphi^2} = 0.$$

If we multiply by $r^2\sin^2\theta$, the first term and the second will depend on r and θ only, the last on φ only. This is impossible unless each of these is a constant, and the constants add to zero. If the last term is $-m^2$, we shall have

$$\frac{d^2\Phi}{d\varphi^2} + m^2\Phi = 0 \tag{2.1}$$

$$\frac{1}{R}\frac{1}{r^2}\frac{d}{dr}\left(r^2\frac{dR}{dr}\right) + \frac{1}{\Theta}\frac{1}{r^2\sin\theta}\frac{d}{d\theta}\left(\sin\theta\frac{d\Theta}{d\theta}\right) - \frac{m^2}{r^2\sin^2\theta} = 0. \tag{2.2}$$

If we multiply by r^2, the first term will depend only on r, the second and third only on θ. Thus again each of these must be a constant. Let the first term be $l(l + 1)$. Then

$$\frac{1}{r^2}\frac{d}{dr}\left(r^2\frac{dR}{dr}\right) - \frac{l(l + 1)}{r^2}R = 0, \tag{2.3}$$

$$\frac{1}{\sin\theta}\frac{d}{d\theta}\left(\sin\theta\frac{d\Theta}{d\theta}\right) + \left[l(l + 1) - \frac{m^2}{\sin^2\theta}\right]\Theta = 0. \tag{2.4}$$

We have now separated the variables, in the sense that we have three ordinary differential equations, (2.3), (2.4), (2.1), for the functions R, Θ, Φ.

3. Spherical Harmonics.—The solution of these equations is not difficult. For (2.3) we may try the assumption $R = r^n$. Then we find that we have a solution if $n(n + 1) = l(l + 1)$, whose solutions are $n = l$, or $n = -(l + 1)$. Thus

$$R = ar^l + \frac{b}{r^{l+1}},$$

where a, b are constants of integration. Equation (2.4) is Legendre's equation. It can be solved by making the substitution

$$\Theta = \sin^m\theta(A_0 + A_1\cos\theta + A_2\cos^2\theta + \cdots).$$

We then find, on substituting in (2.4), that the following relations must exist between the A's:

$$A_0[l(l+1) - m(m+1)] + 2A_2 = 0$$
$$A_1[l(l+1) - (m+1)(m+2)] + 2 \cdot 3A_3 = 0$$
$$A_2[l(l+1) - (m+2)(m+3)] + 3 \cdot 4A_4 = 0$$
$$\cdot \quad \cdot \quad \cdot \quad \cdot \quad \cdot \quad \cdot \quad \cdot \quad \cdot \quad \cdot \quad \cdot \quad \cdot \quad \cdot \quad \cdot \quad \cdot \quad \cdot \quad \cdot \quad \cdot \quad \cdot \quad \cdot$$

These allow us to solve for all the A's in terms of A_0 and A_1, which are constants of integration. In this way we find

$$\begin{aligned}
\Theta = A_0 \sin^m \theta &\left\{ 1 - \frac{[l(l+1) - m(m+1)]}{2!} \cos^2 \theta \right. \\
&+ \frac{[l(l+1) - m(m+1)][l(l+1) - (m+2)(m+3)]}{4!} \cos^4 \theta - \cdots \left.\right\} \\
+ A_1 \sin^m \theta &\left\{ \cos \theta - \frac{[l(l+1) - (m+1)(m+2)]}{3!} \cos^3 \theta \right. \\
&+ \frac{[l(l+1) - (m+1)(m+2)][l(l+1) - (m+3)(m+4)]}{5!} \cos^5 \theta \\
&\qquad\qquad\qquad\qquad\qquad\qquad\qquad\qquad\qquad - \cdots \left.\right\}. \quad (3.1)
\end{aligned}$$

Finally the solution of (2.1) is

$$\Phi = C \sin m\varphi + D \cos m\varphi, \qquad (3.2)$$

where C, D are constants of integration.

We now find that m and l must be chosen to be integers to satisfy certain conditions. Unless m is an integer, an increase of φ by 2π, which brings us back to the same point of space, would lead to a different value of Φ; thus m is an integer, provided that we are solving Laplace's equation in a region that includes all values of φ. Next consider (3.1). If the series do not break off, they can be shown to diverge when $\cos \theta = \pm 1$, or along the axis of coordinates. To show that this is plausible, we find the ratio of the term in $\cos^p \theta$ to that in $\cos^{p-2} \theta$. This ratio is

$$- \frac{[l(l+1) - (m+p-2)(m+p-1)]}{p(p-1)} \cos^2 \theta,$$

which approaches $\cos^2 \theta$ in the limit where p is very large. Thus for $\cos \theta = \pm 1$, this ratio approaches $+1$, so that the terms of the series oscillate without decreasing in magnitude. Under these circumstances it is possible, though not necessary, for a series to diverge. It is also possible for the series to converge, but a closer examination than we shall give shows in the present case that the series actually diverges, and the series would become infinite for $\cos \theta = \pm 1$. If we are solving Laplace's equation in a region including the axis, the potential certainly cannot become infinite along the axis. Thus the series cannot be allowed to diverge. This can be prevented only if l is an integer; for then one of the series (3.1) breaks off to form a polynomial, which is finite for all values of θ. The corresponding polynomials are called "associated Legendre functions," and are commonly denoted by $P_l^m(\cos \theta)$, provided that the constant factor multiplying them is properly chosen. These constant factors, and additional properties of the associated Legendre functions, are taken up in Appendix V. The product of $\sin m\varphi$ or $\cos m\varphi$ by $P_l^m(\cos \theta)$ is called a "spherical harmonic." We then find as our general solution

$$\psi = \left(ar^l + \frac{b}{r^{l+1}}\right) P_l^m(\cos \theta)(C \sin m\varphi + D \cos m\varphi). \qquad (3.3)$$

Here, as with the rectangular case, we can add an infinite number of such terms to get a general solution, and the constants can be chosen so as to fit boundary conditions. For instance, we may solve the problem of the potential between two concentric spheres of radii r_1, r_2, where the potential is given by specified functions of θ, φ on each of the spheres. The method of satisfying these boundary conditions is taken up in Appendix V.

4. Simple Solutions of Laplace's Equation in Spherical Coordinates. We have derived a general solution (3.3) of Laplace's equation. This solution is of great value in complicated problems that can be expressed in terms of spherical coordinates. However, in certain simple cases the solution reduces to very simple forms, which are of great importance. Thus let us ask for a solution corresponding to $l = 0$, $m = 0$. In this case, by (3.1) and (3.2), we see that the functions of θ and φ are constants, so that ψ reduces to $a + b/r$. The first term is a constant of integration, the second the potential of a charge located at the origin. We can determine the constant b if we know the charge, by using Gauss's theorem, or the fact that the potential of a charge q at the origin is $q/4\pi\epsilon_0 r$, so that $b = q/4\pi\epsilon_0$.

The next simplest solution is for $l = 1$, $m = 0$. In this case the first sum of (3.1) fails to break off, and diverges, so that we must set its coefficient equal to zero in our solution; but the second breaks off after the first term, and is a constant times $\cos \theta$. Thus for this case we have

$$\psi = \left(ar + \frac{b}{r^2}\right)\cos \theta. \qquad (4.1)$$

The first term, proportional to $r \cos \theta$, is simply proportional to z, the coordinate along the axis of the spherical coordinates; thus its field is a constant along the z axis, which is surely a solution of Laplace's equation. The next term, proportional to $\cos \theta / r^2$, is particularly interesting: it is the potential of a dipole, which we shall consider in the next section. Before proceeding to it, we note that (4.1) can be used to solve an interesting problem: the potential of a grounded conducting sphere in a uniform external field. Let the external field be along the z axis, and let its magnitude be E. Then at large distances from the sphere we must have $\psi = -Ez = -Er \cos \theta$, or we must have $a = -E$ in (4.1). Next let the radius of the sphere be R, so that the potential must be zero on the sphere. We can arrange to have the potential of (4.1) satisfy this condition, by making

$$aR + \frac{b}{R^2} = 0, \qquad b = ER^3.$$

Thus (4.1) reduces to

$$\psi = \left(-Er + \frac{ER^3}{r^2} \right) \cos \theta, \tag{4.2}$$

which is a solution of Laplace's equation, reduces to zero on the surface of the conducting sphere, and reduces to the correct value at infinite distance.

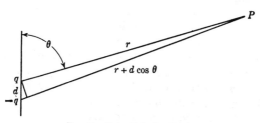

FIG. 8.—Potential of a dipole.

5. The Dipole and the Double Layer.—A dipole by definition consists of two equal and opposite charges $\pm q$ separated by a distance d. The potential at point P is then

$$\frac{q}{4\pi\epsilon_0} \left(\frac{1}{r} - \frac{1}{r + d \cos \theta} \right),$$

where r, θ are as shown in Fig. 8. If d is small, this is approximately

$$\frac{q}{4\pi\epsilon_0} \frac{d \cos \theta}{r^2} = -\frac{qd}{4\pi\epsilon_0} \left[\text{grad} \left(\frac{1}{r} \right) \cdot \mathbf{n} \right],$$

where **n** is the unit vector along the axis of the dipole. The product qd is called the "dipole moment," and is denoted by m. Thus the potential of the dipole is

$$\psi = \frac{m}{4\pi\epsilon_0}\frac{\cos\theta}{r^2}. \qquad (5.1)$$

In speaking of dipoles, it is customary to assume that q is infinitely large, d infinitesimally small, in such a way that the dipole moment m is finite, and (5.1) forms the exact potential. By comparison with (4.1) we see that the second term of that expression is the potential of a dipole of moment $4\pi\epsilon_0 b$, located at the origin.

By analogy with this treatment of the dipole and its potential, one can set up solutions of Laplace's equations similar to (4.1), but corresponding to larger values of l than unity. One can then interpret these solutions as being the potentials of more complicated distributions of charge at the origin. In this way one arrives at the definitions of quadrupoles, and a whole series of higher multipoles, corresponding to successively larger values of l. We consider these higher multipoles in Appendix VI. They are of particular importance in studying the external electric fields (and, as we shall see later, of magnetic fields) of atoms and molecules on the one hand, and of atomic nuclei and their elementary constituents on the other. We often are interested in their fields at a large distance compared with their dimensions, and then they act to a good approximation like multipoles. Thus, for instance, a diatomic molecule like HCl, one of whose atoms (H) tends to acquire a positive charge, the other (Cl) a negative charge, has a field at a distance that consists of a dipole as its leading term. On the other hand, a molecule like N_2, which is symmetrical, has no dipole moment. We notice from (3.3) that the higher l is, the more rapidly the potential falls off with distance, varying as $r^{-(l+1)}$. Thus, since the higher multipoles correspond to higher values of l, their fields fall off more rapidly, and at large distances the field of the multipole of lowest l value predominates. For nuclei, the leading term in the field at large distances, aside from the electrostatic potential varying with $1/r$ resulting from the nuclear charge, is generally a magnetic rather than an electrical term.

The reader might reasonably be surprised that in our present discussion we started by trying to solve Laplace's equation, which is the special case of Poisson's equation which we meet in a region in which there are no charges, and end up by finding the potential resulting from a point charge, dipole, or higher multipole. How, he might ask, did charges get into the theory? The answer is that our spherical coordinate system introduces a singularity at the origin, and our final solutions, although they satisfy Laplace's equation perfectly properly everywhere except at the origin, fail to satisfy it at the origin. The term in (3.3) in r^l has no singularity at the origin, and if we know that there is no charge distribution of any sort at the origin, that term alone must be used in setting up a solution of Laplace's equation. On the other hand, the terms in r^l become infinite at infinite distance from the origin. Closer examination of them shows that they must be connected with the existence of infinite amounts of charge at infinite distance. Thus, to take a very simple case, suppose we have the potential $r \cos \theta$, corresponding to the first term of (4.1). This, as we have seen, is the potential of a uniform field, such as would be set up within an infinite parallel plate condenser. To set it up, we should then have to have equal and opposite charge distributions over two infinitely great condenser plates at infinite distance, and each would carry infinite charge. If, then, by some condition of a problem, we know that there is no charge at infinite distance, we must not use the terms in r^l in the solution (3.3) of Laplace's equation. Of course, if there is no charge at infinity, so that we do not use the terms in r^l, and no charge at the origin, so that we do not use the terms in $r^{-(l+1)}$, and no charge anywhere between, as is implied by the fact that we are trying to solve Laplace's equation, then there will be no field, the potential will be constant everywhere, and there is no problem to be solved.

Returning to the discussion of dipoles, we have seen that (5.1) represents the potential of a dipole of moment m. Similarly the potential found in (4.2), for the problem of a grounded conducting sphere in a uniform external field, is the potential of the superposition of a uniform field, and of a dipole of moment $4\pi\epsilon_0 R^3 E$. We thus see that a conducting sphere of radius R acquires an induced dipole moment of this amount in an external field. When an object thus acquires a dipole moment, it is said to be "polarized," and if the moment is proportional to the field, the ratio is defined as the polarizability α. Thus we have

$$m = \alpha E. \qquad (5.2)$$

As we have just seen, the polarizability of a conducting sphere of radius R is $4\pi\epsilon_0 R^3$.

Often we have occasion to consider, not a single dipole, but a so-called "double layer," a surface distribution of dipoles. Thus let two surfaces a distance d apart carry surface charges $\pm\sigma$ per unit area. Then we may say that the dipole moment per unit area is σd. We note that this double layer is equivalent to a parallel plate condenser, and that the potential difference between the plates is $\sigma d/\epsilon_0$, or the dipole moment per unit area divided by ϵ_0. The field is zero outside the double layer. We note that the normal component of **E** is discontinuous by amount σ/ϵ_0 at a charged surface (as seen in Sec. 1, Chap. II), and the potential is discontinuous by an amount (dipole moment per unit area) divided by ϵ_0 at a double layer.

6. Green's Solution for a Bounded Region.—So far in this chapter we have been considering the direct solution of Laplace's equation, by separation of variables, in rectangular or spherical coordinates. We now consider the other important solution, using Green's method. We proceed as in Chap. II, Sec. 4, only we apply Green's theorem to the volume V between the small sphere of radius R surrounding the point P, and a larger surface S'. Then, substituting $\psi = 1/r$ in Green's theorem, (4.2) of Chap. II, and proceeding as in the derivation of (4.3) of Chap. II, we have in addition a surface integral over S'. We then have in general

$$4\pi\varphi = - \int_V \frac{\nabla^2\varphi}{r}\, dv - \int_{S'} \left(\varphi \operatorname{grad} \frac{1}{r} \cdot \mathbf{n} - \frac{1}{r} \operatorname{grad}\, \varphi \cdot \mathbf{n} \right) d\alpha.$$

This, like (4.3) of Chap. II, is a general theorem holding for any function φ. If in particular $\nabla^2\varphi = -\rho/\epsilon_0$, we have

$$\varphi = \frac{1}{4\pi\epsilon_0} \int_V \frac{\rho}{r}\, dv - \int_{S'} \left[\frac{\varphi}{4\pi}\left(\operatorname{grad} \frac{1}{r} \right) \cdot \mathbf{n} - \frac{1}{4\pi r}\, (\operatorname{grad}\, \varphi \cdot \mathbf{n}) \right] da. \quad (6.1)$$

This expresses φ at any point within V as the sum of two terms: first, the volume integral, representing the potential of the charges within V; secondly, the surface integral, which can be computed if the values of φ and its normal derivative are known at all points of the surface S'. Thus this theorem is in proper form to use in a case in which the behavior of the potential is known around the boundary of a region. Using the results of the preceding section, we can give a physical interpretation to the terms of the surface integral. Suppose we tried to set up such a distribution of surface-charge density, and such a double layer, on S', that the field and potential within S' would have the values actually present in the problem, but so that the field and potential outside S' would be zero. The normal component of field on the inner side of S' would be $-(\operatorname{grad}\, \varphi) \cdot \mathbf{n}$, so that the discontinuity in field at the surface is $(\operatorname{grad}\, \varphi) \cdot \mathbf{n}$. This discontinuity of field would be produced by a surface charge of density $\epsilon_0(\text{grad } \varphi) \cdot \mathbf{n}$. The contribution of this charge to the potential would then be

$$\int \frac{\epsilon_0(\text{grad } \varphi) \cdot \mathbf{n} \, da}{4\pi\epsilon_0 r},$$

which is the second term of the surface integral in (6.1). Similarly the potential within S' is φ, so that the discontinuity is $-\varphi$, and the corresponding dipole moment per unit area to produce this discontinuity is $-\epsilon_0\varphi$. The contribution of this to the potential is

$$\int \frac{\epsilon_0\varphi}{4\pi\epsilon_0} \left(\text{grad } \frac{1}{r}\right) \cdot \mathbf{n} \, da,$$

where the differentiation in the gradient is with respect to the coordinates of the point P. It is the negative of this if the differentiation is with respect to the coordinates x, y, z of the other end of the vector r, as in (6.1). The integral is then just the first term of the surface integral in (6.1). The distribution of surface charge and double layer over S' which we have just discussed is called "Green's distribution." We now see the significance of (6.1): the potential produced by the charge ρ within V, and Green's distribution over S', is just the potential φ that we desire within V, but is zero everywhere outside.

Problems

1. It takes several volts of energy to remove an electron from the interior of a metal to the region outside. Find how many volts, if we represent the surface layer of the metal by a double layer consisting of two parallel sheets of charge; a sheet of negative electricity, of density as if there were electrons of charge 1.60×10^{-19} coulomb, spread out uniformly with a density of one to a square 4×10^{-8} cm on a side; and inside that at a distance of 0.5×10^{-8} cm a similar sheet of positive charge.

2. Find the components of electric field resulting from a dipole, both in rectangular and in spherical coordinates.

3. A charge q is at a distance r from a polarizable dipole, whose moment is α times the field in which it is located, where α is the polarizability. Find the force of attraction between charge and dipole.

4. Referring to Prob. 9, Chap. I, show that the force between a charge q and an uncharged conducting sphere can be found from the result of Prob. 3 above, taken together with the polarizability of a conducting sphere, provided that the distance between charge and sphere is large compared with the radius of the sphere.

5. A dipole of fixed dipole moment is placed in an external electric field. Prove that there is a torque on the dipole proportional to its dipole moment and the magnitude of the electric field, and find how the torque depends on angle.

6. A dipole of fixed dipole moment is placed in an external electric field that is not constant. Prove that there is a force on the dipole depending on the rate of change of field with position, and find how this force depends on the orientation of the dipole and other features of the field.

7. Find the potential as a function of position in a region bounded by surfaces at $x = 0$, $x = L$, $y = 0$, extending to infinity along the y axis, subject to the boundary condition that the potential is zero along the two infinite surfaces $x = 0$, $x = L$, but that it is an arbitrary function of x along the surface from $x = 0$ to $x = L$, $y = 0$. Build up the solution out of individual solutions varying sinusoidally with x, and exponentially with y, noting that they must decrease rather than increase exponentially as y increases.

8. Find the potential as a function of position within a sphere, if the surface of one hemisphere is kept at a potential $+V$, the other hemisphere at a potential $-V$.

9. Set up Laplace's equation in cylindrical coordinates, and solve by separation of variables.

10. A hollow pipe of circular cross section is infinitely long, and is grounded. A disk maintained at potential V practically closes the pipe at a certain point, but is insulated from it. Find the potential as a function of position within the pipe.

11. Show that a solution of Laplace's equation in two dimensions can be written as a linear combination of terms of the form $r^n \genfrac{}{}{0pt}{}{\cos}{\sin} n\theta$, where n can be any positive or negative integer.

12. Given the two-dimensional potential on a circle of radius R about an origin of the form $V(\theta)$ ($0 \leq \theta \leq 2\pi$), where $V(\theta)$ is continuous. Using Green's theorem, show that the potential at any point (r,φ) inside the circle is given by

$$V(r,\varphi) = \frac{R^2 - r^2}{2\pi} \int_0^{2\pi} \frac{V(\theta)\, d\theta}{R^2 + r^2 - 2ar \cos(\varphi - \theta)}.$$

This is known as "Poisson's integral." What form would it take for points exterior to the circle?

13. Find the force and torque on a dipole in the field of a point charge. Find the force exerted by the dipole on the point charge, and verify Newton's third law

CHAPTER IV

DIELECTRICS

A dielectric, or insulator, is a material containing dipoles whose dipole moment is ordinarily proportional to the applied electric intensity. These dipoles can arise physically in two ways. In the first place, the molecules of any material are composed of positive nuclei, surrounded by rapidly moving negatively charged electrons. These electrons move very freely through the atom or molecule; they are prevented from escaping, however, by intense electric fields. When the molecule is placed in an external electric field, the electrons tend to be displaced by the field, much as the free electrons in a metal are displaced. They cannot travel outside the molecule, however; instead, they pile up on the surface of the molecule, resulting effectively in a surface charge, and the net result is that the molecule becomes polarized, or becomes a dipole. The situation is similar to that of the polarization of a conducting sphere in an external field, discussed in Chap. III, Sec. 5. We can even find a value for the polarizability of a molecule, using the formula $\alpha = 4\pi\epsilon_0 R^3$ of that section, and assuming for R a molecular dimension, which is of the right order of magnitude. If there are N atoms or molecules in volume V, so that the number per unit volume is N/V, and if each one acquires a moment αE, there will then be a dipole moment per unit volume of $(N/V)\alpha E$. This moment is called the "polarization," and is denoted by P; it is a vector function of position, the polarization being in the same direction as E, in an isotropic dielectric.

The other mechanism by which dipoles can be set up is found in certain materials containing polar molecules; that is, molecules that have dipole moments even in the absence of an external field. Thus a chemical substance like HCl, containing a positive and a negative ion, has dipole molecules, the H end of the molecule being positively, the Cl end negatively charged. In the absence of an external field, the molecules in gaseous or liquid HCl will be oriented in random directions, so that even though each molecule has a moment, the average moment per unit volume will be zero. An impressed field, however, tends to orient the molecules, and it can be shown that there is a net

41

dipole moment resulting from this, which is proportional to the field. Thus this effect, like the other, gives **P** proportional to **E.** There is a difference, however, which allows the effects to be separated. The induced dipoles are independent of temperature. The orientation of permanent dipoles, however, is opposed by temperature agitation, and by kinetic theory it can be shown that the resulting polarization in a given external field is inversely proportional to the absolute temperature. By investigating the temperature variation of the dipole moment, then, we can experimentally separate the two effects, and can find values both for the polarizability of the individual molecules, and for their permanent dipole moments. For our present purposes, however, we need not distinguish between the two sources of dipoles, and need merely assume the existence of a polarization vector **P,** proportional to **E,** the constant of proportionality of course depending on the nature of the dielectric.

1. The Polarization and the Displacement.—Let us consider a surface S bounding a volume V, within a dielectric. We start with the dielectric unpolarized, and then allow it to polarize. In this process, certain charges will have been carried across the surface S, so that, although originally the volume contains no net charge, there will be such a charge after polarization. Let us find this charge, by computing the amount of charge flowing across an element da of the surface in the process of polarization. We may consider the polarization to consist of many dipoles, each of charge q, with displacement **d** (a vector, pointing along the direction of the dipole moment) between the equal and opposite charges. If **n** is the outer normal to V, the displacement of a charge q along **n** will be $\mathbf{d} \cdot \mathbf{n}$. All those charges contained in the small volume of area da, height $\mathbf{d} \cdot \mathbf{n}$, will be carried over da in the process of polarization. If there are N/V dipoles per unit volume, there will have been a charge of $q(N/V)(\mathbf{d} \cdot \mathbf{n}) \, da$ in this small volume, so that the total charge carried out over da will be $(N/V)(q\mathbf{d} \cdot \mathbf{n}) \, da$. But $(N/V)q\mathbf{d}$ is simply the polarization **P,** the dipole moment per unit volume. Thus the charge carried out over da is $\mathbf{P} \cdot \mathbf{n} \, da$, and the total charge carried out over the whole surface is the surface integral of this quantity.

If ρ' is the resulting charge density within the volume, set up by the displacement of charge, the charge that has been carried out will be $-\int \rho' \, dv$. Thus we have

$$\int_S \mathbf{P} \cdot \mathbf{n} \, da = - \int_V \rho' \, dv. \tag{1.1}$$

Using the divergence theorem, the left side can be rewritten $\int \text{div } \mathbf{P} \, dv$;

since the relation (1.1) must hold for any volume, the integrands must then be equal, or we have

$$\text{div } \mathbf{P} = -\rho'. \tag{1.2}$$

We may now combine this equation with Eq. (3.3), Chap. II, which we may write $\epsilon_0 \text{ div } \mathbf{E} = \rho$, where ρ represents the volume density of charge. We must think carefully how to describe the situation, however. We have two types of charge density: the charge that we deliberately place on our conducting or dielectric bodies, and the charge that automatically appears as a result of polarization. We shall call the first sort the "real charge," the second sort the "polarized charge." Both sorts can produce electric intensity $\mathbf{E}$, so that, in Eq. (3.3), Chap. II, the charge density that appears should be the sum of real and polarized charge.

It is customary, however, to use the symbol ρ to denote merely the real charge, so that for instance ρ will be zero within an ordinary dielectric. Thus we must replace our equation by

$$\epsilon_0 \text{ div } \mathbf{E} = \rho + \rho',$$

which means the same as Eq. (3.3), Chap. II, but is expressed differently because of the different meaning we now assign to ρ. Using (1.2), we then have

$$\epsilon_0 \text{ div } \mathbf{E} = \rho - \text{div } \mathbf{P}, \qquad \text{div } (\epsilon_0 \mathbf{E} + \mathbf{P}) = \rho.$$

The combination $\epsilon_0 \mathbf{E} + \mathbf{P}$ comes into the theory so often and in such an important way that it is given a special name, the electric displacement, and a special symbol, $\mathbf{D}$. Thus we write

$$\text{div } \mathbf{D} = \rho, \qquad \mathbf{D} = \epsilon_0 \mathbf{E} + \mathbf{P}. \tag{1.3}$$

Equation (1.3) is one of the fundamental equations of electromagnetic theory, and is one of Maxwell's equations.

2. The Dielectric Constant.—Since $\mathbf{P}$ is proportional to $\mathbf{E}$, the displacement $\mathbf{D}$ is also proportional to $\mathbf{E}$. We define the ratio of $\mathbf{D}$ to $\mathbf{E}$ as the permittivity, and denote it by ϵ. Thus we have

$$\mathbf{D} = \epsilon \mathbf{E}. \tag{2.1}$$

The ratio of ϵ to ϵ_0, which we may call the permittivity of free space, is the dielectric constant, sometimes called the "specific inductive capacity," and is denoted by κ_e. Thus we may write

$$\kappa_e = \frac{\epsilon}{\epsilon_0}, \qquad \mathbf{D} = \kappa_e \epsilon_0 \mathbf{E}. \tag{2.2}$$

The ratio of polarization $\mathbf{P}$ to $\epsilon_0\mathbf{E}$ is called the "susceptibility," and is denoted by χ_e. Thus we have

$$\mathbf{P} = \chi_e\epsilon_0\mathbf{E}, \qquad \kappa_e = 1 + \chi_e. \tag{2.3}$$

Clearly the susceptibility is proportional to the polarizability of the molecules, and the number of molecules per unit volume: in fact, since $\mathbf{P} = (N/V)\alpha\mathbf{E}$, we have

$$\chi_e = \frac{N}{V}\frac{\alpha}{\epsilon_0}. \tag{2.4}$$

We thus have means for finding the dielectric constant of a material, if we know the polarizability of its molecules. There is one word of caution to be expressed, however. It turns out that the polarizability of a molecule is affected by the presence of neighboring molecules. Thus in (2.4) we cannot use the value of α that would be obtained for a molecule in the absence of neighbors, and expect the resulting dielectric constant to be correct for a dense dielectric. We shall take up this correction later.

3. Boundary Conditions at the Surface of a Dielectric.—There are two fundamental equations of electrostatics: (1.3) and Eq. (3.4) of Chap. I, or

$$\operatorname{div}\mathbf{D} = \rho, \qquad \operatorname{curl}\mathbf{E} = 0. \tag{3.1}$$

These equations take on special forms in the case commonly met in electrostatics, in which the dielectrics consist of materials each with a

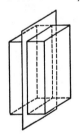

uniform dielectric constant, but with surfaces of discontinuity of dielectric constant between the materials. Within a homogeneous dielectric containing no charges, we have $\operatorname{div}\mathbf{D} = \operatorname{div}\epsilon\mathbf{E} = 0$, so that $\operatorname{div}\mathbf{E} = 0$. From $\operatorname{curl}\mathbf{E} = 0$ we conclude as before that $\mathbf{E} = -\operatorname{grad}\varphi$, and we can then find that $\nabla^2\varphi = 0$, so that Laplace's equation is satisfied for the potential, as in the absence of a dielectric.

At a surface of discontinuity, however, the situation is quite different. There ϵ is a discontinuous function of position, so that we cannot infer from $\operatorname{div}\epsilon\mathbf{E} = 0$ that $\operatorname{div}\mathbf{E} = 0$; it will not be true. Let us instead investigate an uncharged surface of discontinuity between two

Fig. 9.—Gauss's theorem for normal component of D.

dielectrics by Gauss's theorem, using arguments like those in Chap. II, Sec. 1. Over unit area of the surface we erect a thin rectangular volume, as in Fig. 9. We note, from the divergence theorem and the equation $\operatorname{div}\mathbf{D} = \rho$, that Gauss's theorem becomes

$$\int_S \mathbf{D} \cdot \mathbf{n} \, da = \int_V \rho \, dv. \tag{3.2}$$

Since there is no real charge within our volume, this tells us that the total flux of $\mathbf{D}$ out of the volume is zero; that is, that the flux in over one of the faces equals the flux out over the other, which demands that the normal component of $\mathbf{D}$ must be the same on both faces. In other words, we conclude that the normal component of $\mathbf{D}$ is continuous at an uncharged surface of separation between two homogeneous dielectrics.

By an argument similar to that which we have just applied, we may apply Eq. (1.1), rather than (3.2), to the small volume of Fig. 9. Let us suppose that the dielectric is to the left of the surface of separation, empty space to the right. There will then be polarization to the left, and not to the right, so that $\int_S \mathbf{P} \cdot \mathbf{n} \, da$ will be the negative of the component of polarization along the normal pointing away from the dielectric. By (1.1), this equals the polarization charge included within the volume. This must clearly be a surface charge, and since unit area of the surface is included in the volume, we have the result that the normal component of $\mathbf{P}$, pointing out of a dielectric, equals the surface polarization charge that appears on the surface as a result of the polarization. It is of course clear that, since the normal component of $\mathbf{D}$ is continuous at an uncharged surface, and the normal component of $\mathbf{P}$ is not continuous, the normal component of $\mathbf{E}$ must likewise be discontinuous.

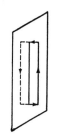

FIG. 10.—Stokes's theorem for surface of discontinuity.

Next we consider the implication of the equation curl $\mathbf{E} = 0$. We draw a contour as shown in Fig. 10, one long side on one side of the surface of separation, the other on the other. By Stokes's theorem in vector analysis, the surface integral of the normal component of the curl of a vector over a surface equals the line integral of the tangential component of the same vector around the contour bounding the surface. Applying this theorem to a surface bounded by the contour in Fig. 10, we note that, since curl $\mathbf{E}$ is zero, the line integral of the tangential component of $\mathbf{E}$ about the contour must be zero, which means that the tangential component of $\mathbf{E}$ must be equal on both long sides of the surface. We then have derived the following general results:

From div $\mathbf{D} = \rho$: normal component of $\mathbf{D}$ continuous
From curl $\mathbf{E} = 0$: tangential component of $\mathbf{E}$ continuous,   (3.3)

where these conditions hold at an uncharged surface of separation between dielectrics.

4. Electrostatic Problems Involving Dielectrics, and the Condenser.—Without the principles derived in this chapter, it is almost impossible to see how to solve electrostatic problems involving dielectrics as well as charges. The dielectrics acquire polarized charges on their surfaces, and if we do not know the amount and distribution of these charges, we cannot use our ordinary electrostatic principles to compute the field. The problem is not unlike that which we met in electrostatic problems involving conductors, where we also did not know the distribution of surface charges. We solved that problem by obtaining a potential that was a solution of Laplace's or Poisson's equation in the region outside conductors, but satisfied certain boundary conditions at the surface of a conductor: the potential had to reduce to an appropriate constant on the surface of each conductor, which implied that the field $\mathbf{E}$ was normal to the surface, or that the tangential component of $\mathbf{E}$ was zero outside this surface. We see that this condition was a special case of (3.3): since $\mathbf{E}$ is zero within a conductor, the tangential component of $\mathbf{E}$ must be zero outside, by continuity. Equation (3.3) gives no information about the normal component at the surface of a conductor, for the surface is not uncharged.

We now notice that a method essentially similar to this one can be used in solving an electrostatic problem involving dielectrics. Inside each dielectric, provided that its dielectric constant is uniform, the potential will satisfy Laplace's or Poisson's equation, where the charge density involved is the real charge, which we assume we know about. At each surface of discontinuity, the conditions (3.3) must be satisfied by the appropriate components of $\mathbf{D}$ and $\mathbf{E}$. Combining with the relation (2.1), $\mathbf{D} = \epsilon\mathbf{E}$, the problem is determined. As a first almost trivial example, consider the field of certain charges embedded in an infinite dielectric of permittivity ϵ. By Gauss's theorem, $\int \mathbf{E} \cdot \mathbf{n}\, da = (1/\epsilon)\int \rho\, dv$. This differs from the corresponding equation for free space only in substituting ϵ for ϵ_0. Thus Green's solution of Poisson's equation will carry through just as in Chap. II, Sec. 4, with the one difference that ϵ will appear in the denominator rather than ϵ_0. Thus the potential, and the field $\mathbf{E}$, of a set of charges in a uniform dielectric, will be just $1/\epsilon_e$ as large as if the same charges were located in empty space. We notice, however, that the displacement $\mathbf{D}$ of a set of charges in a uniform dielectric will be the same as if they were in empty space.

As a next example of electrostatic problems involving dielectrics, consider a condenser filled with a uniform dielectric of permittivity ϵ, rather than with empty space. We can use the same solution of Laplace's equation to represent the potential that we did for the case without dielectrics; the only difference is in the relation between the charge on the plates and the potential difference. For a given voltage between the plates, there will be the same field **E** everywhere in the case with dielectric that there was without it. **D** will, however, be κ_e times as great. By Gauss's theorem, the surface charge at a charged surface will equal D, so that the surface charge on the plates will be κ_e times as great as if the dielectric were absent. The capacity, being the charge divided by the voltage, will then be κ_e times as great as if the condenser had no dielectric but empty space. It is interesting to notice that, at the surface between the dielectric and the conducting plate, there will be not only the real surface charge D, but also be a polarization surface charge, equal to $-P$, or $-(\kappa_e - 1)\epsilon_0 E$. The total charge, real and polarization, on the surface, will then be $\epsilon_0 E$, the same as the total charge for the condenser in the absence of dielectric. The reason a dielectric increases the capacity of a condenser, or the charge that its plates carry for a given voltage, is that a fraction $(\kappa_e - 1)/\kappa_e$ or $1 - 1/\kappa_e$ of the charge is effectively canceled by polarization charge, leaving only the fraction $1/\kappa_e$ for producing a field. Thus, to produce a given field, we must have κ_e times as much charge as in the absence of dielectric.

5. A Charge outside a Semi-infinite Dielectric Slab.—In both the problems we have so far considered, the infinite dielectric and the dielectric filling a condenser, the effect of the dielectric is very simple: the field distribution is just as in empty space, but a given charge produces a field only $1/\kappa_e$ as strong as in empty space. If this situation were always true, the problem of dielectrics would be trivial, and we should not have had to go through all the theory that we have presented in this chapter. We shall now give two examples to show that in general the whole problem is entirely different in the presence of dielectrics from what it is without them. First let us take the problem of a charge q, in empty space, a distance d from an infinite plane surface bounding a semi-infinite dielectric of permittivity ϵ. We can solve this problem by an application of the method of images, similar to that in Chap. I, Sec. 4, but more complicated. In Fig. 11 we shall try to satisfy our conditions by the following assumptions: in the free space, to the left of the surface of separation, we assume that the potential is $q/4\pi\epsilon_0 r_1 - q'/4\pi\epsilon_0 r_2$, where r_1 is the distance from the charge to the point where we are finding the potential, r_2 the distance from the image to the point, and where the second term is the potential of an image charge $-q'$, whose magnitude we must still determine. In the dielectric to the right of the plane, we assume that the potential is $q''/4\pi e_1 r_1$, where q'' is also to be determined. These assumed potentials satisfy the first requirement: in the empty space the potential satisfies Poisson's equation, being determined from the charge q; in the dielectric it is determined by Laplace's equation, the charge producing the field being located outside the dielectric. Clearly we cannot have a term varying inversely as r_2, in the dielectric, for there is no real charge located at the image of q.

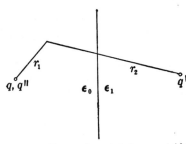

We must now try to satisfy the boundary conditions on the surface. For the tangential component of **E**, we note that the tangential derivative of potential must be the same on both sides of the surface, which means that the potential itself is continuous; this may be used as a substitute for the condition on the continuity of the tangential component of **E**. Thus, remembering that $r_1 = r_2$ at points of the surface, we have

$$q - q' = \frac{q''}{\kappa_e}. \tag{5.1}$$

For the normal component of **D**, we note that the charges q and $-q'$ will produce fields that add rather than subtract. Remembering that we find **D** by taking the gradient of the potential and multiplying by the dielectric constant, this results in

$$q + q' = q''. \tag{5.2}$$

Since (5.1) and (5.2) can be solved for q' and q'' in terms of q, our conditions are compatible, and our assumed potentials give a solution of the problem. Solving for q' and q'', we have

$$q' = \frac{\kappa_e - 1}{\kappa_e + 1}\, q, \qquad q'' = \frac{2\kappa_e}{\kappa_e + 1}\, q. \tag{5.3}$$

Lines of force for this problem are shown in Fig. 12.

We can easily check two special cases of (5.3). First, if $\kappa_e = 1$, so that the whole space is really empty, we have $q' = 0$, $q'' = q$, so

FIG. 11.—Images for point charge outside dielectric slab.

that the field is just that of the charge q in empty space.   Secondly, if κ_e is infinite, $q' = q$, and $q'' = 2q$.   The field $\mathbf{E}$ in the empty space in this case is that produced by the charge q, and an image $-q$; the field within the dielectric is zero, being a field of a finite charge in a medium of infinite dielectric constant.   Thus the field in this case is everywhere the same that we should have for a charge q outside a perfect conductor.   There are as a matter of fact many ways in which a medium of infinite dielectric constant resembles a conductor.

6. Dielectric Sphere in a Uniform Field.—As another example of problems involving dielectrics, we consider a dielectric sphere in a uniform external field, in empty space. Following Eqs. (4.1) and (4.2) of Chap. III, we try the assumption

$$\varphi = \left(-E_0 r + \frac{b}{r^2} \right) \cos \theta$$

outside the sphere, where E_0 is the field at large distances, and where b is to be determined.   This assumption correctly gives the field at large distances, and it satisfies Laplace's equation; if b can be chosen to satisfy the boundary conditions at the surface of the sphere, or

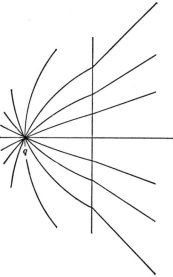

Fig. 12.—Lines of force for point charge outside dielectric slab.

radius R, we shall have a solution of our problem.   Let us also try the assumption that the field is uniform, and equal to E_1, within the sphere.   That is, the potential within the sphere is $-E_1 r \cos \theta$.   As in the preceding problem, we may take as our two boundary conditions the continuity of the potential, and of the normal component of $\mathbf{D}$.   For the potential to be continuous, we must have

$$\left(-E_0 R + \frac{b}{R^2} \right) \cos \theta = -E_1 R \cos \theta, \qquad \text{or} \qquad -E_0 R + \frac{b}{R^2} = -E_1 R.$$

$$(6.1)$$

For the continuity of the normal component of $\mathbf{D}$, we note that the direction of increasing r is the normal direction.   Thus inside the sphere the normal component of $\mathbf{D}$ is $-\epsilon_1 \partial \varphi / \partial r = \epsilon_1 E_1 \cos \theta$, and outside it is $(\epsilon_0 E_0 + 2\epsilon_0 b/R^3) \cos \theta$.   Equating, we have

$$\epsilon_1 E_1 = \epsilon_0 E_0 + \frac{2\epsilon_0 b}{R^3}. \tag{6.2}$$

Solving (6.1) and (6.2) simultaneously, we find

$$E_1 = \frac{3\epsilon_0}{2\epsilon_0 + \epsilon_1} E_0, \qquad 4\pi\epsilon_0 b = 4\pi\epsilon_0 R^3 \frac{(\epsilon_1 - \epsilon_0)}{2\epsilon_0 + \epsilon_1} E_0. \tag{6.3}$$

Thus we have found the complete solution of our problem.   The quantity $4\pi\epsilon_0 b$ is the dipole moment of the dipole that would produce the same field at external points as the polarized dielectric sphere.   It is interesting again to consider the two limiting cases $\epsilon_1 = \epsilon_0$, and $\epsilon_1 = \infty$.   In the first case, we find $E_1 = E_0$, $b = 0$, so that the field is uniform everywhere, as of course it must be if the dielectric sphere is not really present.   In the second case, we find $E_1 = 0$,

$$4\pi\epsilon_0 b = 4\pi\epsilon_0 R^3 E_0,$$

so that again the infinite dielectric constant gives the same fields as for a conductor, as discussed in Chap. III, Sec. 4.

There are several remarks to be made about the solution we have found.   In the first place, there are only very few problems in which we have the simple situation found here, that the field inside the dielectric object is uniform.   This situation holds, in fact, only for an object of ellipsoidal shape.   The general ellipsoid can be solved exactly, as well as the sphere, but by considerably more advanced methods than that used here.   In the second place, our solution holds equally well for the field within an empty spherical hole in the dielectric of permittivity ϵ_1.   As we see from the derivation, to get this result we need only interchange ϵ_0 and ϵ_1 in the solution (6.3).   It is interesting to consider the field E_1 within the cavity.   The ratio E_1/E_0 can be written

$$\frac{E_1}{E_0} = 1 + \frac{\epsilon_1 - \epsilon_0}{2\epsilon_1 + \epsilon_0} = \frac{\epsilon_1}{\epsilon_0}\left[1 - \frac{2(\epsilon_1 - \epsilon_0)}{2\epsilon_1 + \epsilon_0}\right]. \tag{6.4}$$

Thus the field inside the cavity is greater than in the dielectric [since $(\epsilon_1 - \epsilon_0)/(2\epsilon_1 + \epsilon_0)$ is greater than zero], but is less than ϵ_1/ϵ_0 times the field in the dielectric.

7. Field in Flat and Needle-shaped Cavities.—In the preceding section we have found the field within a spherical cavity in a dielectric. It is interesting also to find the field in a disk-shaped cavity whose normal points along the field, and in a needle-shaped cavity pointing along the field.   In the flat cavity, by symmetry, the field inside as well as outside the cavity will be parallel to the field at a large distance, so that, by the continuity of the normal component of **D**, the value of **D** within the cavity will be equal to that in the dielectric.   That is, the value of **E** within the cavity will be ϵ_1/ϵ_0 times the value of **E** in the dielectric, or will be κ_e times the value of **E** in the dielectric.   On the other hand, in a needle-shaped cavity, because of the symmetry, the field will again be everywhere parallel, and because of the continuity of the tangential component of **E**, the value of **E** within the cavity will equal that in the dielectric.

These facts are sometimes used to give definitions of **E** and **D** in a dielectric, in a form that could be made the basis of an experimental method of measuring them: **E** is the force on unit charge in a long needle-shaped cavity parallel to the field, and **D** is ϵ_0 times the force on unit charge in a flat disk-shaped cavity with its normal parallel to the field.   These definitions are analogous to these introduced in a similar way by Lord Kelvin for defining the corresponding magnetic vectors.   It is interesting to note that these two cavities are the limiting forms of ellipsoidal cavities as they get very flat or very elongated, and the solution for the field in an ellipsoidal cavity, which we have mentioned in the preceding section, reduces to these two values in the two limits.   In the intermediate case of the sphere, we have seen in the preceding section that the field in the cavity is intermediate between these two limiting values.

Problems

1. A line charge of linear density σ is placed in a medium of dielectric constant $\kappa_1 = \epsilon_1/\epsilon_0$ parallel to and at a distance a from the plane boundary with another medium of dielectric constant $\kappa_2 = \epsilon_2/\epsilon_0$.   Find the potential in both media, and show that the force per unit length acting on the line charge is given by

$$\frac{\sigma^2}{4\pi\epsilon_1 a} \frac{\kappa_1 - \kappa_2}{\kappa_1 + \kappa_2}.$$

2. A long circular cylinder of radius a and permittivity ϵ is placed with its axis perpendicular to a uniform electric field E_0 in air.   Find the potential inside and outside the cylinder.

3. Find the capacity of a spherical condenser consisting of a conducting sphere of radius r_0, a dielectric of dielectric constant κ_1 for $r_0 < r < r_1$, a dielectric of dielectric constant κ_2 for $r_1 < r < r_2$, and a conducting sphere of radius r_2.

4. Find the capacity per unit length of a cylindrical condenser consisting of a conducting cylinder of radius r_0, a dielectric of dielectric constant κ_1 for $r_0 < r < r_1$, a dielectric of dielectric constant κ_2 for $r_1 < r < r_2$, and a conducting cylinder of radius r_2.

5. Show that, when a line of force cuts through a surface separating two dielectrics of dielectric constants κ_1 and κ_2, it makes angles θ_1 and θ_2 with the normal to the surface in the two media, given by the relation $\kappa_1 \cot \theta_1 = \kappa_2 \cot \theta_2$.

6. A charge q is placed a distance d in front of a semi-infinite dielectric slab of dielectric constant κ_e. Find the force attracting the charge to the slab.

7. A hollow dielectric sphere is placed in a uniform external field. Find the field both inside and outside the sphere, and in the dielectric, in terms of the inner and outer radii of the spherical shell, and its dielectric constant.

CHAPTER V

MAGNETIC FIELDS OF CURRENTS

In Chap. I, we stated that the force on a charge q, moving with a velocity $\mathbf{v}$ in a magnetic field, in which the magnetic induction is $\mathbf{B}$, is $q(\mathbf{v} \times \mathbf{B})$. Similarly we stated that the force on a charge q in an electric field $\mathbf{E}$ was $q\mathbf{E}$. So far we have been studying the electrostatic case. We began by stating the value found experimentally for the electric field $\mathbf{E}$ produced by a charge q. Combination of that statement with the law of force allowed us to get the force between two charges. We shall now begin a similar treatment of magnetic forces. Magnetic fields can be produced, and magnetic forces experienced, by two types of bodies: by charges in motion, or electric currents, and by magnetized bodies, such as permanent magnets. In some cases, the magnetization of magnetic bodies arises from the motion of currents within the atoms, coming from electronic motions, but in other cases it comes from magnetization that is an intrinsic part of the structure of the nuclei or electrons of which the atoms are composed. We could start our study from the fields either of magnets or of currents. We choose to start with currents, as being in a way more fundamental. We shall then show that a current flowing in a small closed path is equivalent to a magnetic dipole, and forms a model for the dipoles present in magnetic materials, which we shall take up in the next chapter.

The magnetic field produced by a steady current, which does not vary with time, is much simpler than that of a current that varies with time, for in the latter case we can have radiation. The problem of magnetic fields of steady currents is called "magnetostatics." It is clear that a single charge in motion cannot produce a static magnetic field, for by its very motion it is found at different points of space at different times. Thus we cannot strictly state the field produced by a charge in motion. Nevertheless, if the charge forms part of a steady current, as if there were a procession of charges, following one after the other in a formation independent of time, then there is a simple law for the field it produces. This law, elucidated by the work of Oersted, of Ampère, and of Biot and Savart, in the early days of the nineteenth century, may be made the basis of the treatment of magnetostatics, just as the law of Coulomb may be made the basis of electrostatics.

1. The Biot-Savart Law.—It is found experimentally that the magnetic induction resulting from a charge q, moving with a velocity **v**, at a distance **r** away from the charge (where **r** is a vector pointing from the charge to the point where the field is being found) is

$$\mathbf{B} = \frac{\mu_0}{4\pi} \frac{q(\mathbf{v} \times \mathbf{r})}{|\mathbf{r}|^3}, \tag{1.1}$$

provided that the moving charge forms part of a current distribution independent of time. In this expression, **B** is the magnetic induction in webers per square meter. Since this is not a familiar unit, we can state that 1 weber/sq m equals a magnetic field of 10^4 gausses, in the more usual units. The quantity μ_0 is

$$\mu_0 = 4\pi \times 10^{-7} \text{ henry/m}, \tag{1.2}$$

where we shall find the physical significance of this number later. The charge q is in coulombs, the velocity in meters per second (a vector), and **r** is in meters. By virtue of the vector product in (1.1), we see that the lines of **B** (that is, lines parallel to the vector **B**) form circles in a plane normal to the velocity vector **v,** with centers at the intersection of that vector and the plane, and in such a sense that a right-hand screw advancing along **v** would rotate in the direction of **B**. If θ is the angle between **v** and **r**, the magnitude of **B** is equal to

$$|\mathbf{B}| = \frac{\mu_0}{4\pi} \frac{q|\mathbf{v}| \sin \theta}{|\mathbf{r}|^2}.$$

We are often interested not in the field of a moving charge but in that of an element of current, as a length ds of wire carrying a current i. We can easily set up an equivalence between the quantity $q\mathbf{v}$ and a corresponding quantity for the current element. Suppose the cross section of the wire is A, and the charge density of charge in the wire is ρ. Assume this charge is moving with an average velocity v. Then the charge crossing any cross section per second is $\rho v A$. On the other hand, if the current flowing is i, measured in amperes, then i must be the charge crossing a cross section per second. Thus we have $\rho v A = i$. If we multiply both sides by ds, the length of the section of wire, we shall have on the left $\rho A \, ds \, v$, which is the product of the charge density, the volume, and the velocity. That is, it is the total charge contained in the element of wire, times its velocity. This then equals $i\,ds$. We may make this into a vector equation by letting $\mathbf{v}$, and $\mathbf{ds}$, both be vectors along the velocity of the charge, or the direction of the wire. Thus we have

$$q\mathbf{v} = i\,\mathbf{ds},$$

so that the magnetic induction resulting from a length $\mathbf{ds}$ of wire carrying a current i is

$$d\mathbf{B} = \frac{\mu_0}{4\pi}\frac{i(\mathbf{ds}\times\mathbf{r})}{|\mathbf{r}|^3}. \tag{1.3}$$

The law described by (1.1) or (1.3) is often called the "Biot-Savart law." As a result of it, and Eq. (1.1) of Chap. I, for the force acting on a moving charge, we can at once find the force between two elements of current, $i_1\mathbf{ds}_1$ and $i_2\mathbf{ds}_2$, flowing in different conductors. This force is

$$d\mathbf{F} = \frac{\mu_0}{4\pi}i_1i_2\frac{[\mathbf{ds}_1\times(\mathbf{ds}_2\times\mathbf{r})]}{|\mathbf{r}|^3}, \tag{1.4}$$

where we have found the force exerted by the second current element on the first, and where $\mathbf{r}$ is the vector from the second to the first. This law (1.4) is the equivalent, for current elements, of Coulomb's law, Eq. (2.2) of Chap. I, expressing the law of force between static charges.

The law of force (1.4) serves essentially to define the ampere, the unit of current. Thus we take two current elements, each of unit length, and at unit distance apart, so oriented that the factor

$$\frac{[\mathbf{ds}_1\times(\mathbf{ds}_2\times\mathbf{r})]}{|\mathbf{r}|^3}$$

in (1.4) equals unity. Then the currents are defined to be an ampere if the resulting force is 10^{-7} newtons. From this it follows that $\mu_0 = 4\pi\times10^{-7}$, as in (1.2), a relation that is now seen to be the result of definition rather than of experiment. This on the other hand is not the case with ϵ_0, which we encountered in Eq. (2.1), Chap. I, and which we found from Coulomb's law. The difference is that we define the unit of current, the ampere, from the Biot-Savart law; we define the coulomb, the unit of charge, as the charge flowing per second in a current of 1 amp; and therefore we are not free to choose the constant in Coulomb's law at will, but must determine it by experiment, with the result given in Eq. (2.1), Chap. I. These relations are discussed in more detail in Appendix II.

2. The Magnetic Field of a Linear and a Circular Current.—We can integrate the Biot-Savart law (1.3) to get the magnetic induction $\mathbf{B}$

resulting from any arbitrary steady current. Naturally in practice we must use such an integrated value to determine the ampere, since the interaction between two elements of current, as considered in (1.4) or in the preceding paragraph, cannot be separated from the forces

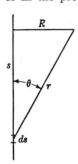

resulting from other current elements in the circuit. First we consider the field of an infinite straight wire carrying a current i. From Fig. 13 we see that $r^2 = R^2 + s^2$, where R is the distance from the wire to the point where we are finding the field, and that $\sin \theta = R/\sqrt{R^2 + s^2}$. Thus we have

$$B = \frac{\mu_0}{4\pi} iR \int_{-\infty}^{\infty} \frac{ds}{(R^2 + s^2)^{3/2}} = \frac{\mu_0 i}{2\pi R}. \qquad (2.1)$$

The direction of **B** is, of course, in circles surrounding the wire.

Fig. 13.—Magnetic field from an infinite straight wire.

Next we consider the field of a circular wire carrying a current, at points along the axis of the wire. The contribution to the field produced by an element ds of the wire, at a point whose distance from the plane is x, will be $(\mu_0/4\pi)i\,ds/(x^2 + R^2)$, as we see in Fig. 14. This field points at an angle to the perpendicular to the plane, however, and only the component along that direction contributes to the resultant field. Thus we must multiply by the factor $R/\sqrt{x^2 + R^2}$ to get the resultant field, which is, after we replace the element ds by the circumference $2\pi R$,

$$B = \frac{\mu_0}{4\pi} i \, \frac{2\pi R^2}{(x^2 + R^2)^{3/2}} = \frac{\mu_0}{2} i \, \frac{R^2}{(x^2 + R^2)^{3/2}}. \qquad (2.2)$$

We can use the solution (2.2) to find the field in the center of a circular current; it is $\mu_0 i/2R$. This solution is sometimes used to provide an experimental determination of the ampere, following the procedure discussed at the end of the preceding sec-

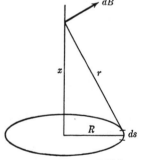

Fig. 14.—Magnetic field from circular current.

tion. It is quite a complicated problem to find the field off the axis in the case of a circular wire.

The solution (2.2) can be used to find the field along the axis of an infinite solenoid, carrying n turns of wire per unit length. To find this we need merely multiply (2.2) by $n\,dx$, and integrate over x, taking account of the turns of the solenoid located at all distances x

from a given point on the axis.   Thus we have

$$B = \frac{\mu_0}{2} ni \int_{-\infty}^{\infty} \frac{R^2}{(x^2 + R^2)^{3/2}} dx = \mu_0 ni. \tag{2.3}$$

We shall shortly prove that this field is found everywhere inside the infinite solenoid, not merely along the axis.

3. The Divergence of B, and the Scalar Potential.—In our study of electrostatics, we first found simple methods of obtaining the field by integrating the contributions resulting from all point charges according to Coulomb's law.   We soon found, however, that this was a method of restricted usefulness.   The really powerful tools of electrostatics come only when we apply general analytical methods, introducing the potential, Gauss's theorem, Poisson's equation, and similar relations.   Here too we shall introduce general analytical methods.   As a first step, we investigate the divergence and the curl of **B**.   We shall first show that div **B** is always zero; then we shall show that curl **B** = 0 in important cases, in which we can therefore introduce a potential, but that this is not true in general, so that we have to adopt other ways of writing **B** in terms of a potential, leading to the concept of a vector potential.

We first consider the divergence of **B**.   The lines of **B** form closed circles around the axis of a current element, and since the divergence of a vector signifies the starting or stopping of the vector, we should expect that div **B** = 0.   We can prove this, directly from (1.1) or (1.3), and if the theorem holds for the field of an element of current, it must hold for the sum or integral of such fields, or for the field of a complete circuit.   We can prove our result by use of the vector relation div (**a** × **b**) = **b** · curl **a** − **a** · curl **b**, where **a** and **b** are any vectors.   We let **a** = **v**, **b** = **r**/|**r**|³, in (1.1).   Since **v**, or **a**, is a constant, the first term will be zero.   Then we have

$$\text{div } \mathbf{B} = \frac{\mu_0}{4\pi} q \left(-\mathbf{v} \cdot \text{curl } \frac{\mathbf{r}}{|\mathbf{r}|^3} \right). \tag{3.1}$$

But the vector **r**/|**r**|³ is simply a vector in the direction of **r**, of magnitude $1/r^2$; that is, it is just like the field of a point charge in electrostatics, whose curl has been proved to be zero in earlier chapters. Thus, from (3.1), we have

$$\text{div } \mathbf{B} = 0. \tag{3.2}$$

Equation (3.2) is one of the fundamental equations of electromagnetic theory, one of Maxwell's equations, which we shall not have to alter in our later development.

Next we consider the curl of **B**. If current flows steadily in a closed loop of wire, we can show that the curl of the resulting magnetic induction is zero everywhere outside the wire, and hence that we can introduce a potential to describe the problem. We shall prove this theorem by setting up a simple and general expression for the potential. As a first step, we consider the solid angle Ω intercepted

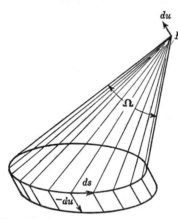

by the loop of wire, at the point P where we are finding the magnetic induction **B**. If we make a displacement **du** of P, as shown in Fig. 15, the solid angle will change by an amount $d\Omega$. This is the same as the change $d\Omega$ produced by a displacement $-$**du** of the loop. If we make the latter displacement, the change of solid angle will be the sum of all the elementary changes of solid angle intercepted by the small parallelograms bounded by the vectors **du,** **ds,** as shown in the figure. The projection of one of these parallelograms on the normal to the

FIG. 15.—Scalar potential of loop of wire carrying current.

vector r is $-(\text{d}\mathbf{u} \times \text{d}\mathbf{s}) \cdot \mathbf{r}/|\mathbf{r}|$, so that the corresponding increment of solid angle is $-(\text{d}\mathbf{u} \times \text{d}\mathbf{s}) \cdot \mathbf{r}/|\mathbf{r}|^3 = -\text{d}\mathbf{u} \cdot (\text{d}\mathbf{s} \times \mathbf{r})/|\mathbf{r}|^3$.

Thus the whole change of solid angle, integrating around the current loop, is

$$d\Omega = -\text{d}\mathbf{u} \cdot \int \frac{\text{d}\mathbf{s} \times \mathbf{r}}{|\mathbf{r}|^3}.$$

But at the same time the change $d\Omega$ must be grad $\Omega \cdot$ **du**. Thus we have

$$-\text{ grad } \Omega = \int \frac{\text{d}\mathbf{s} \times \mathbf{r}}{|\mathbf{r}|^3}.$$

Comparing with (1.3), we see that we have

$$\mathbf{B} = -\frac{\mu_0}{4\pi} i \text{ grad } \Omega. \tag{3.3}$$

In other words. the function $(\mu_0/4\pi)i\Omega$ forms a potential, whose negative gradient gives the magnetic induction **B**. This is generally called a "scalar potential," since it is a scalar quantity, to distinguish it from the vector potential, which we shall soon introduce. It is clear that, in cases in which **B** can be derived from a scalar potential, we must have curl **B** = 0, since the curl of any gradient is zero.

4. The Magnetic Dipole.—Suppose we have a current i flowing in a positive direction around a very small loop of area A, the normal to the loop being **n**. If θ is the angle between **n** and the vector from the loop to a point P, the solid angle subtended by A at P will be $A \cos \theta/r^2$, so that the scalar potential will be

$$\frac{\mu_0 i A}{4\pi} \frac{\cos \theta}{r^2}.$$

FIG. 16.—Surface made up of small current loops.

This expression depends on position just like the potential (5.1) of Chap. III for an electric dipole. Thus a small loop carrying a current has a magnetic induction **B**, whose lines of force are like those of an electric dipole. We call such a small loop a "magnetic dipole," and define the product iA as the dipole moment. (Some writers define the dipole moment as $\mu_0 i A$ instead, with corresponding change in subsequent formulas involving dipole moments.) If this moment is called m, we have

$$\text{Potential of magnetic dipole} = \frac{\mu_0 m}{4\pi} \frac{\cos \theta}{r^2}. \tag{4.1}$$

In terms of this formulation, there is an interesting way in which we can regard the magnetic field of a current loop as coming from a distribution of magnetic dipoles over a surface spanning the loop. In Fig. 16, we may divide the surface into many small loops, letting a current i flow in each of them, in a positive direction. These currents will cancel each other on all the interior boundaries of the loops, but not over the outer boundary. Thus the sum of the currents of all the small loops will give just the current originally present in the large loop, so that the magnetic field of the small loops must equal that of the large loop. Each of the small loops, of area A, however, produces a field like a dipole of moment iA. In other words, a distribution of dipoles over the surface, or a double layer, whose dipole moment per unit area is i, distributed over any surface spanning the current loop, will have the same magnetic induction at all points that the current loop itself has.

5. Ampère's Law.—Although **B** is derivable from a potential function, we must not assume without further study that the line integral of **B** about a closed contour is always zero, which we should be inclined to deduce by analogy with Sec. 3, Chap. I. Let us compute the line integral of the tangential component of **B** about a curve as shown in Fig. 17, enclosing the current. Using (3.3), we have

$$\int \mathbf{B} \cdot \mathbf{ds} = -\left(\frac{\mu_0 i}{4\pi}\right) \int d\Omega.$$

If we make a small excursion, and return to our starting point, Ω comes back to its initial value, $\int d\Omega$ is zero, and the line integral $\int \mathbf{B} \cdot \mathbf{ds}$ is zero. If, however, as in Fig. 17, our path encloses the current, the situation is quite different. If we start integrating when we are in the plane of the loop, Ω will be 2π, corresponding to a hemisphere. As we traverse the path in the direction shown, Ω will

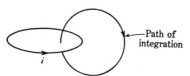

Fig. 17.—Path of integration around current loop.

decrease from this value to zero, become negative, and when we return to our original point it will be -2π. This is on the assumption that solid angles are positive when the point P is above a plane in which the current is circulating in a positive direction, as in Fig. 15. The net change in Ω is then -4π, so that $\int \mathbf{B} \cdot \mathbf{ds} = \mu_0 i$. In other words, when we integrate the tangential component of **B** about a contour enclosing the current i, the integral is $\mu_0 i$; whereas when we integrate about a contour enclosing no current, the integral is zero. The statement of these facts is Ampère's law: the line integral of **B** about a closed contour, in a region containing steady currents, equals μ_0 times the total current flowing through the contour of integration. This general case obviously follows from the derivation we have given, if we regard the whole current flow as being made up of many loops, some of which thread through our contour of integration, some of which do not.

Ampère's law can often be used, as Gauss's theorem was used in electrostatics, to get the answers to simple problems in the magnetic fields of currents. For instance, we may solve for the field of a linear current, taken up in Sec. 2. If we apply the theorem to a circular contour surrounding the current, **B** is parallel to **ds** by symmetry, and is constant around a circular contour. Thus $B(2\pi R) = \mu_0 i$,

$$B = \left(\frac{\mu_0 i}{2\pi R}\right),$$

as found by direct integration in (2.1). Similarly we can use the theorem to find the field inside a solenoid. Setting up a contour as shown in Fig. 18, we may assume that the field points along the axis by symmetry, and that it is zero outside the solenoid. Then, if the length of the contour is unity, the line integral $\int \mathbf{B} \cdot \mathbf{ds}$ is simply the value of $\mathbf{B}$ inside the solenoid. The current enclosed by the contour is ni, if there are n turns in unit length, each carrying current i. Thus we have $B = \mu_0 ni$, as in (2.3). Our present answer is more general, however, for it holds for any point within the solenoid, not merely for points on the axis.

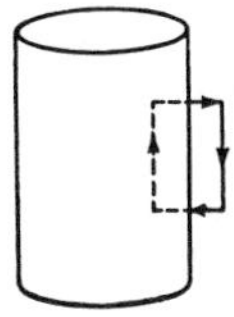

Ampère's law, which may be stated in the integral form as

$$\int \mathbf{B} \cdot \mathbf{ds} = \mu_0 \Sigma i, \qquad (5.1)$$

FIG. 18.— Ampère's law for a solenoid.

where Σi indicates the total current threading through the contour, may be written in a differential form by using Stokes's theorem. This theorem of vector analysis states that

$$\int \mathbf{A} \cdot \mathbf{ds} = \int \text{curl } \mathbf{A} \cdot \mathbf{n} \, da,$$

where $\mathbf{A}$ is any vector function of position. That is, the line integral of the tangential component of $\mathbf{A}$ about a closed contour equals the surface integral of the normal component of curl $\mathbf{A}$ over any surface spanning the contour. Using this theorem, we can transform the left side of (5.1). To transform the right side, we introduce the current-density vector $\mathbf{J}$. This represents the number of amperes per square meter flowing in a continuous conductor. The current flowing across a surface element da is then $\mathbf{J} \cdot \mathbf{n} \, da$, so that the total current flowing through the contour is $\int \mathbf{J} \cdot \mathbf{n} \, da$ integrated over a surface spanning the contour. Thus we have as a result of (5.1)

$$\int \text{curl } \mathbf{B} \cdot \mathbf{n} \, da = \mu_0 \int \mathbf{J} \cdot \mathbf{n} \, da.$$

This result must hold for any contour and any surface spanning it; thus the integrands must be equal, and since the integrands are equal for any direction of the vector $\mathbf{n}$, the vectors themselves must be equal, or

$$\text{curl } \mathbf{B} = \mu_0 \mathbf{J}. \qquad (5.2)$$

This is the differential formulation of Ampère's law.

We see from (5.2) that, in a region containing no current flow, curl $\mathbf{B} = 0$, but this does not hold within a conductor. In empty space where curl $\mathbf{B} = 0$, we can introduce a potential, as we have done in (3.3), but this cannot be done inside a conductor carrying a current.

The potential we introduced in Sec. 3 for a single loop of current was multiple-valued; that is, as we integrated **B** around a contour enclosing the current the potential increased by $\mu_0 i$, so that if we integrated n times around the contour we should increase the potential by $n\mu_0 i$. In other words, for any single value of the potential, all values obtained by adding or subtracting integral multiples of $\mu_0 i$ are equally legitimate. If now we have many current loops in our space, we can change the potential by integral multiples of μ_0 times any one of the currents, by traversing suitable contours. With enough current loops, this means that we can obtain almost any value of the potential we desire, at a given point of space; and if the current is distributed over the volume, so that we can enclose any amount by traversing a suitable contour within the volume, we can obtain any value of potential whatever at a given point of space. In this case the usefulness of the idea of potential breaks down completely. The integral of **B** is not independent of path, and there is no unique way of defining a scalar potential.

6. The Vector Potential.—It is clear from our discussion of the preceding section that a scalar potential is not useful for discussing a magnetic field, except in the special case where the current flows in a single loop. Mathematically, we have seen in (5.2) that curl **B** is not in general zero, so that the conditions for introducing a potential do not exist. There is, however, another quite different way of setting up a potential function, which is much more general, and equally useful. This is to set up what is called a "vector potential." The fundamental characteristic of a potential is that it is a function that we differentiate to get the vector field we are interested in. The reason why the potential is a useful device in electrostatics is that it is easy to compute it from the charge distribution, by the solution of Poisson's equation given in Eq. (3.5) of Chap. I.

The vector-potential solution for the magnetic field of currents has the same useful features. The existence of a vector potential is based on Eq. (3.2), div **B** = 0, satisfied by the magnetic induction. By a well-known theorem of vector analysis, div curl **A** = 0, where **A** is any arbitrary vector; that is, the divergence of any curl is zero. It seems reasonable to assume from this that we can write

$$\mathbf{B} = \text{curl } \mathbf{A}, \tag{6.1}$$

where **A** is a vector function of position, which is the vector potential we have spoken of. By making the assumption (6.1), (3.2) is automatically satisfied, and we must choose **A** so as to satisfy (5.2) as well. This gives us

$$\text{curl curl } \mathbf{A} = \mu_0 \mathbf{J}.$$

By a theorem of vector analysis, curl curl $\mathbf{A}$ = grad div $\mathbf{A} - \nabla^2\mathbf{A}$, where $\mathbf{A}$ is an arbitrary vector. We shall now assume that

$$\text{div } \mathbf{A} = 0. \tag{6.2}$$

We are allowed to make this assumption; it turns out to be the case that, to determine a vector function uniquely, we must specify both its divergence and its curl at all points of space, and (6.1) leaves the divergence undetermined. Making the assumption (6.2), we then have

$$\nabla^2\mathbf{A} = -\mu_0\mathbf{J}. \tag{6.3}$$

That is, $\mathbf{A}$ satisfies Poisson's equation, as φ was shown to in Eq. (3.4), Chap. II. Solving by Eq. (4.3), Chap. II, for an unbounded region, we have the general solution

$$\mathbf{A} = \frac{\mu_0}{4\pi} \int \frac{\mathbf{J}}{r} \, dv. \tag{6.4}$$

A solution similar to Eq. (6.1) of Chap. III, for a bounded region, can be set up as well. In (6.4) we have a general solution for the vector potential, and hence for the magnetic induction, of any known distribution of currents.

We may show from (6.4) that we are in fact led to the same solution from the Biot-Savart law which we have already discussed, just as the general solution of Poisson's equation led in the electrostatic case to the same result as an elementary discussion of Coulomb's law. In the first place, suppose the current density $\mathbf{J}$ is flowing in a length ds of a conductor of area a. Then $Ja = i$, the current flowing in the conductor, and $J \, dv = Ja \, ds = i \, ds$. Thus the vector potential arising from a current element $i \, \mathbf{ds}$, where we regard $\mathbf{ds}$ as a vector, is

$$d\mathbf{A} = \frac{\mu_0}{4\pi} \frac{i \, \mathbf{ds}}{r}, \tag{6.5}$$

a vector in the direction of the current element. To find the value of $\mathbf{B}$ arising from this vector potential, we take its curl. Noting that $\mathbf{ds}$ is a constant vector, r a scalar function of position, we need to use the formula curl $(f\mathbf{F}) = f$ curl $\mathbf{F}$ + grad $f \times \mathbf{F}$, where f is a scalar, $\mathbf{F}$ a vector function. In our case $\mathbf{F}$ is constant, so that we have

$$d\mathbf{B} = \frac{\mu_0}{4\pi} i \text{ grad} \left(\frac{1}{r}\right) \times \mathbf{ds}.$$

But grad $(1/r) = -1/r^2$ grad r, and grad $r = \mathbf{r}/|\mathbf{r}|$. Thus finally

$$d\mathbf{B} = \frac{\mu_0}{4\pi} i \frac{\mathbf{ds} \times \mathbf{r}}{|\mathbf{r}|^3},$$

in agreement with (1.3). We have, then, given a proof of the Biot-Savart law from Eqs. (3.2) and (5.2), and have illustrated our method of handling the vector potential, but have not arrived at a new way of solving the magnetic problem. The solution for a bounded region, similar to Eq. (6.1) of Chap. III, would be new, but it is not of enough practical importance for us to work it out. Although the vector potential has given us no new information in this case, we shall find that in problems involving time variation of charge and current we can still use it, and it then proves a very valuable method of solving for the magnetic induction.

Problems

1. A steady current flows in a circular loop of wire, of radius R. Find the vector potential of the resulting magnetic induction, at large distances compared with R, by adding the contributions to the vector potential due to the separate elements of current.

2. Compute the field, from the potential of the preceding problem, and show that it is approximately that resulting from a single dipole. Find the strength of the dipole, in terms of current and radius R, and check the value derived in the text using the scalar potential.

3. Two parallel straight wires carry equal currents. Work out the magnetic induction due to the two together, in the two cases where the currents flow in the same or in opposite directions, drawing diagrams of the lines of force.

4. Find the magnetic induction at points inside a wire carrying a current, assuming that the wire is straight and of circular cross section and that the current has constant density throughout the wire.

5. Set up the equation for the vector potential in empty space, in cylindrical coordinates, finding components of $\mathbf{A}$ along r, θ, and z. (*Hint:* Set up the equations curl curl $\mathbf{A} = 0$, div $\mathbf{A} = 0$, using derivatives of the latter, to help simplify the terms.)

6. Use the result of Prob. 5 to find the vector potential of an infinitely long wire carrying a constant current i. From this vector potential find the value of $\mathbf{B}$.

7. As in Prob. 5, set up the equation for the vector potential in spherical polar coordinates. Use this result to find the vector potential of a magnetic dipole, comparing with the result of Prob. 1.

8. Two long parallel conductors of circular cross section, each of radius b, are separated by a distance $2a$ between their axes of rotation. If they carry equal and opposite steady currents i, find expressions for the vector potential at all points of a plane (the xy plane) perpendicular to the direction of the currents. From this obtain the magnetic field at interior points of either conductor and set up the integral for the force per unit length with which they repel each other. (Use as an origin the point midway between the wires in the xy plane.) Carry the integration as far as you can.

9. Show that, for any current distribution parallel to a fixed line (the z axis), the magnetic induction may be obtained from a vector potential that has only a z component. From this prove that the lines of magnetic force are given by the equation $A(q_1,q_2) = $ constant, where A is the z component of the vector potential, and q_1, q_2 are orthogonal coordinates in a plane normal to the z axis. Apply this to the results of Prob. 8, and construct a field plot for this case.

CHAPTER VI

MAGNETIC MATERIALS

Just as a dielectric contains electric dipoles that contribute to the field, so there are magnetic media that contain magnetic dipoles. These media are of three sorts: diamagnetic, paramagnetic, and ferromagnetic. A diamagnetic medium contains no permanent dipoles, but only dipoles that are induced by an external field. The atoms of a substance contain electrons that are free in a sense to move about inside the atom, somewhat like the charge in a perfect conductor. We have used this property of the electrons in describing the electric polarization of the atom in an electric field. Similarly in a region where the magnetic induction changes with time, currents are induced in the atom, following the general law of electromagnetic induction which we shall take up in the next chapter. These currents circulate about the direction of the magnetic induction, and thus produce magnetic moments, which prove to be in such a direction that they oppose the field already present. A medium that possesses only this dipole moment opposite to **B** is called a "diamagnetic medium."

Some media contain permanent dipoles. In an unmagnetized body, the dipoles are oriented at random, just as permanent electric dipoles are oriented at random in the absence of an external field, but under the action of an external field the dipoles are oriented, resulting in a dipole moment which, as in the corresponding electric case that we discussed in Chap. IV, is proportional to the external field, and inversely proportional to the absolute temperature. This effect is called "paramagnetism." When it exists, it is usually great enough to mask the diamagnetism which is always present, and which has the opposite sign. The permanent magnetic moments responsible for paramagnetism arise in two ways. In the first place, the theory of atomic structure shows that in many cases there is a permanent circulating current of electrons in the atoms, resulting in a magnetic dipole, as shown in Chap. V. Such circulating currents were first postulated by Ampère, and are often called "Amperian currents," but the explanation of their magnitude and nature was first given by the quantum theory. The diamagnetic atoms are those in which some electrons are circulating in one direction, some in the other, in such a way that their dipole moments cancel. Secondly, the quantum theory shows that an individual electron possesses a magnetic moment, which is just as characteristic a property as its electric charge. In certain ways this moment can be ascribed to a rotation of the charge about an axis, like the rotation of the earth about its axis, resulting thus in a circulating current; but this simple picture of it cannot be entirely justified. In any case, any atom that has unbalanced moments of the spinning electrons will be paramagnetic; but many atoms that contain even numbers of electrons can have equal numbers of electrons oriented in opposite directions, so that the net magnetic moment is zero, and the atoms are diamagnetic, both orbital and spin magnetic moments canceling.

The final type of medium, the ferromagnetic medium, is really an extreme case of paramagnetism. If the permanent dipoles, generally those resulting from electron spin, are very close together in the medium, there proves to be an effect, explainable only on the quantum theory, and called "exchange," which results in a strong tendency for the spins of adjacent atoms or molecules to line up parallel to each other, even in the absence of an external field. Such a parallel orientation can extend, in an unmagnetized body, over volumes of a considerable scale on an atomic order of magnitude, though a small volume by ordinary standards. Such a volume is called a "domain," and an ordinary unmagnetized ferromagnetic body consists of many domains, each with a strong permanent moment, but oriented in different directions. In the presence of an external magnetic field, the domains change the orientation of their permanent moments, lining them up with the external field, until finally with a very large external field the moment reaches a limit when all moments are parallel. This limit is called the "saturation moment." Reversing the field reverses the moments, but there is an effect similar to friction, hindering this reorientation, so that, by the time the external field is reduced to zero, there can still be a considerable moment. This is the origin of permanent magnetism. If the external field is reversed alternately between one direction and then the other, the moment lags behind the field, resulting in the phenomenon of hysteresis.

These properties of ferromagnetic bodies are very complicated, when one tries to investigate them in detail, in contrast to the diamagnetic and paramagnetic bodies, in which the moment is proportional to the field. The ferromagnetic effect decreases with temperature, the individual domains losing their moments at a critical temperature called the "Curie temperature." Above that temperature the body becomes paramagnetic, but the moment, instead of being proportional to $1/T$, where T is the absolute temperature, is proportional to $1/(T - \theta)$, where θ is the Curie temperature. The physical explanation of this decrease of ferromagnetism as the temperature increases is that the tendency toward orientation of the magnetic moments which lines them up at low temperatures is opposed by thermal agitation at high temperatures.

1. The Magnetization Vector.—In Sec. 4, Chap. V, we defined the dipole moment of a magnetic dipole: if a current i circulates in a loop of area A, the moment is iA. In a magnetic medium, we shall have dipoles distributed through the volume of the material, and we shall define the magnetization vector **M** as the vector sum of the dipole moments in unit volume. In a diamagnetic or paramagnetic medium, **M** will be proportional to **B**, the constant of proportionality being negative for diamagnetism, positive for paramagnetism; in a ferromagnetic medium, the relationship between **M** and **B** will be much more complicated.

We now observe that the existence of a magnetization vector within a medium implies the existence of currents. Suppose as in Fig. 19 that we have a small rectangular volume, with magnetization along one of the axes. If there is uniform magnetization within the volume, we may replace the magnetic effect of the volume by a current circulating about its faces, as shown. If the area normal to the magnetization is A, the height h, then the total dipole moment is MAh. This would be produced by a current Mh circulating about an area A. In other

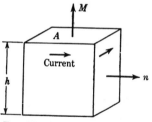

Fig. 19.—Magnetization surface current.

words, we must assume a surface current at the surface of the magnetized volume, numerically equal to the magnetization. We may indicate the direction as well as the magnitude of this surface current density by the vector equation

$$\text{Surface current} = \mathbf{M} \times \mathbf{n}, \tag{1.1}$$

where **n** is the outer normal. This is the surface current that appears at a surface of discontinuity between a region with magnetization **M**, and an unmagnetized region.

It is clear that, more generally, there will be a surface current $(\mathbf{M}' - \mathbf{M}'') \times \mathbf{n}$ at a surface of discontinuity between a region where the magnetization is $\mathbf{M'}$, and one where it is $\mathbf{M''}$, $\mathbf{n}$ being the normal pointing from the region $\mathbf{M'}$ to that where the magnetization is $\mathbf{M''}$. Passing to the limit, there will be a current flowing throughout a volume, when the magnetization varies continuously from point to point within a medium. Let us find the resulting current density. In Fig. 20, we show two adjacent volumes, the upper one being displaced a distance dy along the y axis from the lower one. Let

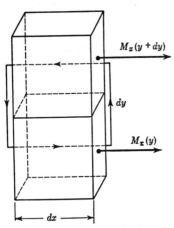

the magnetic moments, in the x direction, be $M'' = M_x(y + dy)$, $M' = M_x(y)$, respectively. Then, considering the contour drawn, the current flowing in it can be written as $J'_z \, dx \, dy$, where $\mathbf{J'}$ is the current density. Thus by the result above we have

$$[M_x(y + dy) - M_x(y)] \, dx = -J'_z \, dx \, dy.$$

Passing to the limit as dx, dy become small, this becomes

$$J'_z = - \frac{\partial M_x}{\partial y}.$$

Fig. 20.—Volume magnetization current.

We should also have a component of J' in the z direction if M_y varied with x; by a similar argument we should find a contribution $\partial M_y/\partial x$. Thus combining terms we have

$$J'_z = \frac{\partial M_y}{\partial x} - \frac{\partial M_x}{\partial y}.$$

But this is merely the z component of the vector equation

$$\mathbf{J'} = \operatorname{curl} \mathbf{M}, \tag{1.2}$$

which is the general relation giving the volume density resulting from a variation of $\mathbf{M}$ from point to point, a relation of which (1.1) is the limiting form at a surface of discontinuity between a magnetized and an unmagnetized region.

2. The Magnetic Field.—The current density $\mathbf{J'}$ which we have written in (1.2) is a current density that is necessarily present whenever we have a magnetized medium. In our introductory remarks we have seen its interpretation, in terms of the flow of electrons within atoms, and of electron spins. It is quite different from the current density $\mathbf{J}$ met for instance in Chap. V, Sec. 5, which was a current density resulting from the ordinary flow of current in conductors. The situation is similar to that which we met in Chap. IV, Sec. 1, where we found two types of charge: what we called the "real charge," the charge we placed on the plates of condensers and other conductors, and the "polarized charge," which automatically appeared in dielectrics. There we denoted the polarized charge by a prime, and here similarly we are denoting the current density $\mathbf{J}'$ resulting from magnetic polarization by a prime. We may, if we choose, refer to $\mathbf{J}$ as the real current (that which flows in ordinary conductors), and to $\mathbf{J}'$ as the polarized current, or the magnetization current.

It now follows from our whole train of argument that the magnetization current is just as effective in producing a magnetic induction as a real current. In other words, in Ampère's law, Eq. (5.2) of Chap. V, we must replace $\mathbf{J}$, the current density, by $\mathbf{J} + \mathbf{J}'$, the sum of real and polarized current density. Hence we may rewrite that law

$$\operatorname{curl} \mathbf{B} = \mu_0(\mathbf{J} + \mathbf{J}') = \mu_0(\mathbf{J} + \operatorname{curl} \mathbf{M}),$$

from which we deduce at once the relation

$$\operatorname{curl}\left(\frac{\mathbf{B}}{\mu_0} - \mathbf{M}\right) = \mathbf{J}. \tag{2.1}$$

The quantity $\mathbf{B}/\mu_0 - \mathbf{M}$ which appears in (2.1) is a sufficiently important quantity so that we give it a name and a symbol: we shall call it the "magnetic field," or "magnetic intensity," and denote it by $\mathbf{H}$. The process of introducing it is essentially similar to that used in Chap. IV, Sec. 1, in introducing the electric displacement $\mathbf{D}$. We may rewrite the definition of $\mathbf{H}$ in the form

$$\mathbf{B} = \mu_0(\mathbf{H} + \mathbf{M}). \tag{2.2}$$

In terms of it, we then have

$$\operatorname{curl} \mathbf{H} = \mathbf{J} \tag{2.3}$$

as the equivalent of (2.1).

In the diamagnetic or paramagnetic medium, we have already stated that $\mathbf{M}$ is proportional to $\mathbf{B}$. Thus both these vectors are also proportional to $\mathbf{H}$, and we may write

$$\mathbf{B} = \mu\mathbf{H} = \kappa_m\mu_0\mathbf{H}, \tag{2.4}$$

where the dimensionless quantity κ_m is generally referred to as the "permeability." The ratio of magnetization to magnetic intensity is generally called the "magnetic susceptibility":

$$\mathbf{M} = \chi_m \mathbf{H}, \qquad \kappa_m = 1 + \chi_m. \tag{2.5}$$

These equations are closely analogous to Eqs. (2.3) of Chap. IV, though μ_0 does not appear in the same way in which ϵ_0 did in the earlier case. Unlike the case of dielectrics, χ_m can be negative (for diamagnetic media) as well as positive (for paramagnetic media). The magnitudes are also quite different: for diamagnetic and paramagnetic media, κ_m differs by very small amounts from unity, and χ_m is very small compared with unity, whereas the dielectric constant κ_e can be quite large compared with unity. For ferromagnetic media, where $\mathbf{M}$ is not proportional to $\mathbf{B}$ or $\mathbf{H}$, we can define a permeability by (2.4), but it will be a function of $\mathbf{H}$, rather than a constant. It is more convenient for such media simply to give experimental curves for $\mathbf{B}$ as a function of $\mathbf{H}$, as is done in the conventional hysteresis curves, which we shall mention in a later section.

The magnetization vector measures the density of magnetic dipoles per unit volume, just as the polarization $\mathbf{P}$ measures the density of electric dipoles. In Chap. IV, we found that $\mathbf{P}$ was associated with a volume and surface-charge density: by Eq. (1.2) of that chapter we found that the volume density ρ' was given by div $\mathbf{P} = -\rho'$, and in Sec. 3 of that chapter we found that at a surface of a dielectric a surface-charge density appeared, equal numerically to the component of $\mathbf{P}$ along the outer normal. In an analogous way we may define a volume density of magnetic poles, equal to $-$ div $\mathbf{M}$, and a surface density equal to the component of $\mathbf{M}$ along the outer normal to a magnetized body. Remembering that div $\mathbf{B} = 0$, we see that div $\mathbf{H} = -$ div $\mathbf{M}$. Thus lines of $\mathbf{H}$ will diverge outward from regions where the density of poles is positive, or from north poles, and will converge to regions of negative density, or south poles. Lines of $\mathbf{B}$, on the other hand, are continuous, on account of the equation div $\mathbf{B} = 0$. We shall shortly see an example of this behavior of magnetic lines, in the case of the magnetized sphere. It should be understood that, although we have an analogy between electric charge and magnetic pole strength, this analogy is far from complete, since electric charges can be separated from each other, whereas magnetic poles can exist only in dipoles, or in polarized bodies.

3. Magnetostatic Problems Involving Magnetic Media.—We have found two fundamental equations dealing with the magnetic field,

$$\text{div } \mathbf{B} = 0, \qquad \text{curl } \mathbf{H} = \mathbf{J}, \tag{3.1}$$

where the first is Eq. (3.2) of Chap. V, and the second (2.3). These, taken together with (2.4), or the corresponding relation for ferromagnetic media given by the hysteresis curve, give enough information so that we can solve magnetostatic problems involving magnetic media as well as currents. Problems in diamagnetism and paramagnetism can be solved as easily as those involving dielectrics, and by essentially the same methods. We generally have a number of media, each of uniform permeability, with surfaces of separation between. Generally also we do not have any currents $\mathbf{J}$ flowing in the magnetic media. Thus inside these media we have div $\mathbf{B} = 0$, from which, since $\mathbf{B} = \mu\mathbf{H}$, and μ is constant, we have div $\mathbf{H} = 0$; furthermore curl $\mathbf{H} = 0$; $\mathbf{H}$ can then be derived from a scalar potential, which satisfies Laplace's equation because of div $\mathbf{H} = 0$. Thus the fields inside the various media can be found by familiar methods, and we have only to satisfy boundary conditions at the surfaces of discontinuity, and to ensure that our fields behave properly at the currents $\mathbf{J}$. The boundary conditions, by methods already used, are

From div $\mathbf{B} = 0$: normal component of $\mathbf{B}$ continuous

From curl $\mathbf{H} = \mathbf{J}$: tangential component of $\mathbf{H}$ continuous at
a surface carrying no current; if there is
a surface current, surface-current density
= (discontinuity in $\mathbf{H}$) $\times$ $\mathbf{n}$.   (3.2)

As a simple example, let us consider the problem of a solenoid filled with a diamagnetic or paramagnetic material. As in Chap. V, Sec. 5, we have a solenoid with n turns per unit length, each carrying a current i. The solenoid is filled with a material of permeability κ_m. The surface-current density is ni; thus the discontinuity in tangential component of $\mathbf{H}$ between the inside and outside of the solenoid is ni. Since $\mathbf{H}$ is zero outside, we see that inside $\mathbf{H} = ni$, and $\mathbf{B} = \mu ni$. In other words, the value of $\mathbf{H}$ within an infinite solenoid is the same, independent of the medium within it, so long as there are the same number of ampere-turns in the winding, but the value of $\mathbf{B}$ is proportional to the permeability of the medium. This problem is one that shows us the proper units to use for measuring $\mathbf{H}$: since $\mathbf{H} = ni$, we measure magnetic intensity in ampere-turns per meter. A problem similar to the solenoid, and conveniently realizable experimentally, is the toroid, a ring-shaped piece of magnetic material, wound round and round with windings, so that the magnetic lines circulate within the ring. Simple application of Ampère's law shows that, just as in the solenoid, the value of $\mathbf{H}$ is equal to the number of ampere-turns per meter in the winding, provided that the cross section is small compared with the length of the toroid, and the value of **B** is μ times as great.

Another simple problem is the field produced by any distribution of currents in an infinite space filled with a medium of permeability κ_m. Because of the equation curl **H** = **J**, the magnetic intensity **H** produced by the currents will be independent of κ_m. Thus **B** will be κ_m times as great as in the corresponding problem in empty space. Since the force on another current element is proportional to **B,** we see that the force between two current elements immersed in a medium of permeability κ_m is κ_m times as great as if the same current elements were in empty space. We may take account of this, in Eq. (1.4), Chap. V, for the force between two current elements, by replacing the quantity μ_0 which appears in that expression by μ. We must remember, however, just as in the corresponding case of dielectrics, that the general case of a number of different magnetic media with surfaces of separation between them is a complicated problem, and that it is by no means true that the value of **B** at any point is merely κ_m times as great as it would be if no magnetic media were present.

4. Uniformly Magnetized Sphere in an External Field.—Just as in the corresponding dielectric problem taken up in Sec. 6, Chap. IV, the problem of a uniformly magnetized sphere in an external field is one that can be solved exactly, and that is of considerable physical importance. Let us first consider a sphere of radius R with a uniform magnetization **M** along the axis of spherical coordinates, without an external field (such as we might have from a permanent magnet). We can later superpose a constant external field. We can easily find that the field of this sphere is as follows:

Inside the sphere, $\mathbf{H} = -\dfrac{\mathbf{M}}{3}, \qquad \mathbf{B} = \dfrac{2}{3}\mu_0\mathbf{M}$, along axis

Outside the sphere, $\mathbf{H} = -\operatorname{grad}\left(\dfrac{MR^3}{3r^2}\cos\theta\right), \qquad \mathbf{B} = \mu_0\mathbf{H}.$ (4.1)

We can verify this solution as follows: As in Sec. 6, Chap. IV, **B** and **H** outside the sphere are the field of a dipole, of moment $\frac{4}{3}\pi R^3 M$ (where we use Chap. V, Sec. 4, to get the moment from the scalar potential). This is simply the moment of a sphere of radius R, with a constant density of magnetization **M**. The field outside the sphere is then a gradient of a scalar potential, so that it is a solution of the equations governing **B** and **H** in empty space. Inside the sphere, the constant values of **B** and **H** are likewise solutions of these equations. We can find the normal components of **B,** and tangential components of **H,** inside and outside the sphere, at radius R, and show them to be continuous. Finally, inside the sphere we have the relation

$$\mathbf{B} = \mu_0(\mathbf{H} + \mathbf{M}),$$

satisfying Eq. (2.2).

The lines of **B** inside and outside the sphere are then as given in Fig. 21. Because of the relation div **B** $= 0$, the lines are closed, the normal flux being continuous at the surface. On the other hand,

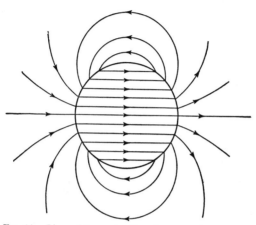

Fig. 21.—Lines of force of uniformly magnetized sphere.

although the lines of **H** outside the sphere are like those of **B,** the lines of **H** inside run in the opposite direction. Thus lines of **H** originate from the right face of the sphere (the north pole of the magnet), and terminate at the left face (the south pole). If now the magnet is in a constant external field **H₀**, whose corresponding value of **B** is **B₀**, pointing along the axis, and equal to $\mu_0\mathbf{H}_0$, we need only add these constant values **H₀** and **B₀** to the solution of (4.1), both inside and outside the sphere, to get the complete solutions. We thus find that inside a magnetized sphere placed in an external field **H₀**, the magnetic field is only $\mathbf{H}_0 - \mathbf{M}/3$. This effect, by which there is a term $-\mathbf{M}/3$ subtracted from **H₀** because of the magnetization, is called the "demagnetizing effect," and the factor ⅓, which depends on the geometry of the sphere, is called the "demagnetizing factor."

We can now use these relations, together with a hysteresis curve, to investigate the magnetic moment that a ferromagnetic sphere would acquire in a given external field. If B, H, represent the values within the sphere, in an external field H_0, we then have

$$H = H_0 - \frac{M}{3}, \qquad \frac{B}{\mu_0} = H_0 + \frac{2M}{3}, \qquad \frac{B}{\mu_0} = -2H + 3H_0. \quad (4.2)$$

If we know the relation between B and H characteristic of the material (that is, the hysteresis curve), we can then find the intersection of this curve with the curve $B/\mu_0 = -2H + 3H_0$, from (4.2). This intersection will give us the values of B and H inside the sphere in any external field. From those, we can get M, and hence the mag-

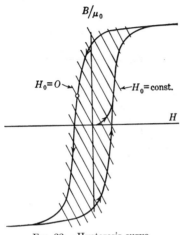

netic moment of the sphere. An example of a hysteresis curve, with the straight lines (4.2) superposed, is given in Fig. 22. Conversely, by a measurement of the magnetic moment, we can work backward and find the hysteresis curve.

As an example of this method, we may ask what are the values of B and H within the sphere, if we first magnetize to saturation, and then remove the external field. If $H_0 = 0$, we have $B/\mu_0 = -2H$, and the values of B and H are determined by the intersection shown in the figure. We note that this has a much smaller value of B

FIG. 22.—Hysteresis curve.

(and hence a much smaller magnetic moment) than if the demagnetizing factor were smaller. For suppose the factor is L, so that

$$H = H_0 - LM,$$

$B/\mu_0 = H_0 + (1 - L)M.$ Then we have

$$\frac{B}{\mu_0} = H\left(1 - \frac{1}{L}\right) + \frac{H_0}{L}. \quad (4.3)$$

Thus, as L approaches zero, the slope of the straight line corresponding to $H_0 = 0$ becomes negatively infinite, or the line coincides with the axis of ordinates in the BH diagram, so that the B inside the permanently magnetized object becomes much greater.

This situation can be approached in a practical case. It turns out that we can obtain an exact solution of the problem of a uniformly magnetized body, similar to (4.1), not only for the sphere, but for the ellipsoid; though the solutions are much more complicated, involving elliptic integrals. When the ellipsoid is infinitely elongated along the axis, the magnet approaches an infinitely thin bar magnet; in this case the demagnetizing factor approaches zero. The opposite limit is the flattened disk-shaped ellipsoid, in which the demagnetizing factor approaches unity, so that B/μ_0 approaches H_0, and in the absence of an external field there is no B within the magnet. Because of the possibility of getting exact solutions, accurate experiments on the magnetic behavior of magnetic materials are generally made on samples of an ellipsoidal shape. As a final example of similar arguments, we can find the values of B and H within ellipsoidal cavities in a magnetized medium, as we did in Sec. 7, Chap. IV, for the corresponding dielectric case. We find that with a long thin cavity the value of H equals the value H_0 in the medium at a large distance from the cavity, whereas with a disk-shaped cavity the value of H equals the value of B_0 in the medium at a large distance, divided by μ_0.

5. Magnetomotive Force.—Practical magnets are not made in the spherical or ellipsoidal shapes discussed in the preceding section, but rather in the shape of a closed ring of some shape with an air gap, and the external field is supplied by windings surrounding the ring, as shown in Fig. 23. In this case, if the permeability of the magnetic medium is high, the lines of force will mostly flow through the magnetic material except near the gap, where the lines of force will partly cross the gap where we are trying to produce a high field, and will partly leak around the sides of that gap. It is almost impossible to calculate accurately the form of the resulting field; we do not have the simplifying feature that the

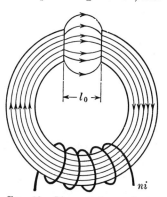

FIG. 23.—Lines of force of ring magnet.

field is constant within the magnetic medium, as we do with the ellipsoidal or spherical magnet. We can, however, make certain assumptions that allow us to use a certain amount of theory in discussing such a magnet.

Let us assume that the leakage is concentrated in a relatively short part of the magnet, near the poles. Then, if the cross section is uniform through most of the length, we note that the flux of **B** must be the same through most cross sections (because of the lack of leakage) and hence **B** must be constant along the length of the magnet (because of the uniform cross section). Let this constant value of **B** within the magnet be B_1, and the corresponding value of **H** be H_1. Furthermore let the length of the magnetic material, along a mean circumference, be l_1. In the gap, if there were no leakage, the value of **B** would be the same as inside the magnet; because of leakage, however, it will be smaller. Suppose the value of **B** in the gap is then $B_0 = \alpha B_1$, where α is a factor less than unity, which is sometimes called the "leakage factor." Then the value of **H** in the gap will be

$$H_0 = \frac{B_0}{\mu_0} = \frac{\alpha B_1}{\mu_0}.$$

Let the length of the gap be l_0. Now we apply Ampère's law to a contour through the middle of the magnet. From (2.3), Ampère's law may be stated in the form

$$\int \mathbf{H} \cdot \mathbf{ds} = ni, \tag{5.1}$$

where there are n turns of wire, each carrying i amp, threading the contour. We then have approximately

$$H_0 l_0 + H_1 l_1 = ni. \tag{5.2}$$

We may then rewrite (5.2) in the form

$$\frac{B_1}{\mu_0} = -\frac{l_1}{\alpha l_0} H_1 + \frac{ni}{\alpha l_0}. \tag{5.3}$$

This equation, relating B_1 and H_1, is essentially equivalent to (4.3). In place of the factor $(1/L - 1)$ multiplying H we have $l_1/\alpha l_0$; in place of H_0 we have the term in ni. It is customary to call the number ni, which is the number of ampere-turns, the "magnetomotive force" (mmf), by analogy with the electromotive force $\int \mathbf{E} \cdot \mathbf{ds}$ met in the electric field. As in the preceding section, in a plot of B_1 versus H_1 a line of constant mmf will be a straight line of negative slope. In an actual case, l_1 will ordinarily be large compared with l_0, and α will be rather small. Thus the case resembles that of a small demagnetizing factor, which we have discussed in connection with (4.3), so that the value of B_1 inside the magnet will be large if the mmf is removed, as in a permanent magnet.

We may note a number of interesting consequences of our theory. If the magnet is operated as an electromagnet, with large mmf, the value of B_1 will be inversely proportional to l_0; that is, for a large gap, we shall have to have a correspondingly high mmf to get saturation of the ferromagnetic material, which we presumably want to do to operate the magnet efficiently. A small value of α, on the contrary, allows us to saturate the material with a smaller mmf, but it gives correspondingly smaller fields in the gap. If the magnet is operated as a permanent magnet, the field in the gap, which we are primarily interested in, is given by

$$H_0 = - \frac{l_1}{l_0} H_1,$$

so that, if the material is used with a given value of B_1 and of H_1, the field in the gap will be inversely proportional to the length of the gap, for constant magnet length l_1. It can be shown, from arguments of energy that we are not yet prepared to understand, that a permanent magnet can be made with least material if it is made with those particular values of B_1 and H_1 which make the product B_1H_1 a maximum; if this condition is applied, the dimensions of a magnet designed for a given field and given gap are determined.

Problems

1. An electron of charge e, mass m, rotates in a circle of radius r with velocity v. Using classical methods, show that the magnetic moment is proportional to the angular momentum. If the angular momentum is $h/2\pi$, where h is Planck's constant, find the magnetic moment, in mks units. If we make a simple model of a ferromagnetic substance by putting one such moment at each point of a simple cubic lattice of lattice spacing 3 A, find the magnetization per unit volume, M, when all moments are parallel.

2. A sphere of radius a carrying a uniform surface-charge density σ is rotated about a diameter with constant angular velocity ω. Calculate the magnitude and direction of the magnetic intensity $\mathbf{H}$ at the center of the sphere. Show that this value of $\mathbf{H}$ is the same at all points inside the sphere.

3. A long circular cylinder of permeability κ_m is placed in a uniform external magnetic field that is perpendicular to the generators of the cylinder, resulting in a uniform magnetization of the cylinder. Find expressions for $\mathbf{B}$ and $\mathbf{H}$ inside and outside the cylinder, and the demagnetizing factor.

4. A long straight wire carrying a steady current i is placed parallel to and at a distance a from a large thick plane slab of permeability κ_m. Show that the field inside the slab is that which would be produced by a current $2i/(\kappa_m + 1)$ in the same wire embedded in an infinite medium of permeability κ_m. Show further that the field in the space in front of the slab is that produced by a current i and another current $i(\kappa_m - 1)/(\kappa_m + 1)$ parallel to and at a distance a behind the interface, both in empty space. What is the force per unit length acting on the current-carrying wire?

5. Discuss the magnetic field of permanent magnets, showing that $\mathbf{H}$ may be obtained as the negative gradient of a single-valued scalar potential ψ, which satisfies Poisson's equation $\nabla^2\psi = -\rho'$, with $\rho' = \operatorname{div} \mathbf{H}$. If one defines magnetic-pole strength as $p = \int \rho' \, dv$, show that two concentrated poles exert mechanical forces on each other given by $F = \frac{p_1 p_2 \kappa_m \mu_0}{4\pi r^2}$, if κ_m is the permeability of the medium in which they are embedded.

CHAPTER VII

ELECTROMAGNETIC INDUCTION
AND MAXWELL'S EQUATIONS

The history of electromagnetism has shown as its most conspicuous feature the gradual discovery of interconnections among problems that were at first supposed to be separated. The two oldest fields were electrostatics, and the magnetism of permanent magnets and of ferromagnetic bodies. Early in the nineteenth century Oersted and others demonstrated the magnetic effects of continuous currents, bringing together the study of the electric current, which developed with the discovery of various forms of batteries in the eighteenth century, and the study of magnetism. Faraday, soon after this work, began looking for a converse effect. He reasoned that, if currents could produce magnetism, magnets should be able to produce currents. His first idea was simple, but wrong. He wound two coils of wire together, but insulated from each other, and planned to pass current through one of them, converting it into an electromagnet. He hoped that this magnet, with its lines of induction threading through the other coil, would cause a continuous current to flow through that coil, just as a continuous current produces a continuous magnetic field. His experiment did not show such a current; even though his battery was powerful enough so that his primary coil was heated red-hot, still no current flowed in the secondary. But he was a good enough observer to notice that, though there was no steady current in the secondary, there was nevertheless a transient current when the current was started in the primary, and a transient in the opposite direction when the primary current was interrupted. This suggested to him that the effect he was seeking really existed, but that the induced current for which he was searching was proportional, not to the magnetic flux itself, but to its time rate of change. This law of electromagnetic induction, which bears Faraday's name, is the foundation of the study of electromagnetic theory.

Faraday thought, not in mathematical language, but in terms of lines of force. We have already seen how this concept, leading to the field theory of electrostatics and magnetostatics, allows us to formulate problems involving dielectrics, conductors, and magnetic materials in a form that is much more powerful than any method based on the concept of action at a distance. In electromagnetic induction as well the idea of lines of force, and of flux, proved to be of the greatest value, the statement of Faraday's law involving the time rate of change of magnetic flux through the coil. The proper mathematical formulation of the methods of lines of force did not come, however, until the work of Maxwell, a number of years later. Maxwell brought together the various laws of electrostatics and magnetostatics which we have already discussed, and correlated them with Faraday's law, expressed in a similar differential or vector form. He was able to see that the system of laws so built up was mathematically inconsistent, and that to make it consistent he had to introduce a certain quantity called the "displacement current," which was in principle susceptible of experimental observation, but which had not been observed at the time. By combining these laws, he set up the equations known as Maxwell's equations, which have been the foundation of the theory of electromagnetism ever since. These equations did not merely describe the phenomena known at Maxwell's time; they predicted the existence of electromagnetic waves, which later proved to be identical with light waves, thus enormously extending the range of the theory. These advances we shall take up in later chapters; we now consider the detailed nature of the law of electromagnetic induction, and its role in electromagnetic theory.

1. The Law of Electromagnetic Induction.—It is found experimentally that, when the flux of magnetic induction **B** through a wire or other conducting circuit changes with time, a current flows in the circuit. The law of electromagnetic induction can be stated most simply, not in terms of the current set up in the circuit, but in terms of the emf leading to the current. By definition, the emf around a circuit equals the total work done, both by electric and magnetic forces and by any other sort of forces, such as those concerned in chemical processes, per unit charge, in carrying a charge around the circuit. Faraday's law states that the emf induced in a circuit equals the negative of the rate of increase of flux of **B** through the circuit. The flux of **B** is by definition the surface integral of the normal component of **B** over a surface whose perimeter is the circuit in question. We may then write Faraday's law in integral form in the following way:

$$\int \mathbf{E} \cdot \mathbf{ds} = -\frac{d}{dt} \int \mathbf{B} \cdot \mathbf{n} \, da. \tag{1.1}$$

The term on the left is the emf; in the case of electromagnetic induction, the force acting on the charge is, as we shall soon see, an electric force. The term on the right is the time rate of decrease of magnetic induction through the circuit. The surface integral is to be extended over any surface spanning the circuit. Because div $\mathbf{B} = 0$, this integral will be the same over any such surface; for if we take two such surfaces, the total flux of $\mathbf{B}$ out of the volume enclosed by the two surfaces must be zero, so that the flux through each surface must be the same. In (1.1), the emf is to be expressed in volts, and the flux in webers; we note that 1 volt must be equivalent to 1 weber/sec.

It is a familiar fact associated with electromagnetic induction that there are several ways in which the magnetic flux through a circuit can change. First, the circuit itself may move from one part of the magnetic field to another. This is the process used in the ordinary electromagnetic generator, or dynamo. Secondly, the circuit may be stationary, but the magnetic induction may be a function of time, either because it is produced by currents that are varying with time, or because magnetic materials such as permanent magnets are moving. If $\mathbf{B}$ is changing with time, but the circuit is fixed, we may take the time derivative in (1.1) inside the integral sign. Also we may transform the left side of (1.1) into a surface integral, by using Stokes's theorem. Thus we have

$$\int \text{curl } \mathbf{E} \cdot \mathbf{n} \, da = - \int \left(\frac{\partial \mathbf{B}}{\partial t} \right) \cdot \mathbf{n} \, da.$$

This equation must hold for a circuit of any shape, and with any direction of the normal $\mathbf{n}$; thus the integrands on the two sides must be equal, or we have

$$\text{curl } \mathbf{E} = - \frac{\partial \mathbf{B}}{\partial t}. \tag{1.2}$$

This equation is the differential form of Faraday's law, and as we shall see later it is one of the fundamental equations of electromagnetic theory, one of Maxwell's equations.

2. Self- and Mutual Induction.—If a current i_1 flows in a circuit, it produces a magnetic field $\mathbf{B}_1$ proportional to i_1 at all points of space. If i_1 changes slowly, the magnetic field is very closely the same at any instant as it would be if the current had a constant value equal to its instantaneous value; such a condition is known as a "quasistationary state." There will then be a time rate of change of $\mathbf{B}_1$ proportional to the time rate of change of i_1. The flux of $\mathbf{B}_1$ through the circuit itself will then vary as i_1 does, so that we shall have an emf in this circuit proportional to di_1/dt, which we may write as

$$\text{emf}_1 = -L \frac{di_1}{dt},$$

where L is the coefficient of self-induction. Similarly the flux of $\mathbf{B}_1$ through another circuit will vary as i_1 does, so that the emf in a second circuit is

$$\text{emf}_2 = -M \frac{di_1}{dt},$$

where M is the coefficient of mutual induction.

We can easily set up general expressions for the coefficients of self- and mutual induction by use of Eq. (6.5) of Chap. V, expressing the contribution to the vector potential resulting from a current element $i\,\mathbf{ds}$. From the discussion above, L equals the flux of $\mathbf{B}$ through a circuit per unit current flowing in it, and M the flux of $\mathbf{B}$ through a circuit per unit current flowing in another circuit. But remembering that $\mathbf{B} = \text{curl } \mathbf{A}$, we have

$$\int \mathbf{B} \cdot \mathbf{n}\, da = \int \text{curl } \mathbf{A} \cdot \mathbf{n}\, da = \int \mathbf{A} \cdot \mathbf{ds},$$

in which the last line integral follows from Stokes's theorem. Using Eqs. (6.4) and (6.5), Chap. V, which are

$$\mathbf{dA} = \frac{\mu_0}{4\pi} \frac{\mathbf{J}}{r}\, dv = \frac{\mu_0}{4\pi} \frac{i\, \mathbf{ds}}{r} \tag{2.1}$$

we have

$$L \text{ or } M = \frac{\mu_0}{4\pi} \int \int \frac{\mathbf{J} \cdot \mathbf{J}'}{ii'r}\, dv\, dv' \tag{2.2}$$

Here dv and dv' are two elements of volume, and r is the distance between them. The integral is a double integral, since each of the elements of length is integrated around the contour. For L, both dv and dv' are to be integrated around the single circuit 1; for M, dv is to be integrated around circuit 1, dv' around circuit 2. Because of the symmetrical formula, we see that the mutual inductance between two circuits is the same, no matter which of them we regard as the source of the current, and which as the one in which the emf is being induced.

In many special cases, we can make direct calculations of the coefficients of self- and mutual inductance, by simpler methods than the use of (2.2), which in practice usually involves complicated integrations. For, if we can find the magnetic flux resulting from the system of currents in the circuits, we can often integrate these fluxes over the circuits directly. Thus, for instance, consider the self-inductance per unit length of two parallel wires of radius a, at distance D apart between their centers, shown by Fig. 7, which was used earlier in computing the capacity of the same parallel wires. To find the magnetic field produced by a unit current flowing up through the right-hand wire, down through the left-hand wire, we first consider unit current flowing up through an infinitely thin conductor located at point 2 of the figure, and an equal current flowing down through a linear conductor at point 1. We note an analogy between the electrostatic and magnetic problems: the scalar potential φ in the electrostatic case, for a distribution of charge along these lines, is given by Eq. (3.5) of Chap. I, and the vector potential **A** in the magnetic case for a distribution of current is given by Eq. (6.4) of Chap. V. These formulas are the same, except for two differences: the formula for **A** has μ_0 in the numerator in place of ϵ_0 in the denominator, and the formula for **A** involves current density in place of charge density, and hence results in a vector.

Since every current element is in the same direction in our present problem, **A** in this case will also be in the same direction, normal to the plane of the paper in the figure. If **k** is a unit vector along this direction, and A is the magnitude of **A**, we have $\mathbf{A} = \mathbf{k}A$. To take the curl of **A**, and find **B**, we use the rule of vector analysis that

$$\text{curl } f\mathbf{F} = f \text{ curl } \mathbf{F} + \text{grad } f \times \mathbf{F},$$

where f is a scalar, **F** a vector. Letting f stand for A, **F** for **k**, and remembering that **F** is then a constant vector, we have

$$\mathbf{B} = \text{curl } \mathbf{A} = \text{grad } A \times \mathbf{k}.$$

If on the other hand we had had unit charge along the lines instead of unit current, the electric field would have been

$$\mathbf{E} = - \text{grad } \varphi,$$

where as we have just seen φ equals $A/\epsilon_0\mu_0$. Thus we see that the **B** resulting from unit current, and the **E** resulting from unit charge, are related by the equation

$$\mathbf{B} = \epsilon_0\mu_0(\mathbf{k} \times \mathbf{E}).$$

The magnitudes, in other words, are proportional, and the vectors are at right angles to each other, so that the lines of magnetic force are identical with the electric equipotentials in the problem of Fig. 7. In particular, the circular boundaries of the wires are lines of magnetic force.

Now, to integrate the normal component of **B** over a surface spanning the circuit, we may integrate over a plane surface including the two lines 1 and 2. The normal component of $\mathbf{B}$ in this case is just $\epsilon_0\mu_0$ times the tangential component of $\mathbf{E}$ in the corresponding electrical case. Thus the surface integral of the normal component of $\mathbf{B}$ over unit length of the transmission line, integrating between the boundaries of the two cylinders, is $\epsilon_0\mu_0$ times the corresponding difference of potential in the electrical case. The inductance is the flux of $\mathbf{B}$ per unit current; thus it is $\epsilon_0\mu_0$ times the difference of potential per unit charge, or is $\epsilon_0\mu_0$ divided by the electrostatic capacity per unit length. Using Eq. (2.1) of Chap. II for this capacity, we then find that the inductance per unit length is

$$L = \frac{\mu_0}{\pi} \cosh^{-1} \frac{D}{2a}. \tag{2.3}$$

In this calculation, we have neglected magnetic flux through the interior of the wires. This flux would be small in any case, and could almost be neglected. If the wires, instead of being solid, were hollow, so that the current flowed in a thin shell on the surface, and there was no flux inside, we should have just the condition for which (2.3) was correct. The method of calculation we have used here, leading to a simple relationship between problems in electrostatics and magnetism, can often be used, and can be justified in a much more general case than the one we have considered here.

3. The Displacement Current.—In the course of our work, we have derived four fundamental electromagnetic equations, (1.3) of Chap. IV, (3.2) of Chap. V, (3.1) of Chap. VI, and (1.2) of the present chapter, or

$$\begin{aligned}
\operatorname{div} \mathbf{D} &= \rho, \\
\operatorname{div} \mathbf{B} &= 0, \\
\operatorname{curl} \mathbf{E} &= -\frac{\partial \mathbf{B}}{\partial t}, \\
\operatorname{curl} \mathbf{H} &= \mathbf{J}.
\end{aligned} \tag{3.1}$$

These are almost Maxwell's equations, but there is a difficulty with the last of them. We have derived it from Ampère's law, on the basis of steady closed currents, and for this case it is correct. The difficulty occurs when we try to apply the equation to nonstationary cases. Suppose we have a current flowing in an open circuit, as in the discharge of a condenser. The current starts at the positively charged plate, whose charge diminishes as the current flows to the negatively charged plate and annuls the charge there. Thus we can look upon the condenser plates as sources or sinks of current. Now, if we take the divergence of the last equation, we have

$$\text{div curl } \mathbf{H} = \text{div } \mathbf{J},$$

and, since the divergence of any curl is zero, we find that div **J** equals zero, which means that the current is always closed and there are no sources or sinks. Thus we are led to a contradiction.

Maxwell concluded from this that the last equation of (3.1) must be incomplete, and that to the term **J** must be added another term, such that the sum of the two had no divergence. We can easily find what this term must be, from the equation of continuity for the flow of current. The divergence of **J** measures the flux outward over the surface of unit volume. If there is current flowing outward, the charge within unit volume must be decreasing, the flux equaling the rate of decrease of charge in unit volume. Thus we have

$$\text{div } \mathbf{J} = -\frac{\partial \rho}{\partial t}. \tag{3.2}$$

We may rewrite this equation of continuity, using the equation div $\mathbf{D} = \rho$, and obtaining

$$\text{div} \left(\mathbf{J} + \frac{\partial \mathbf{D}}{\partial t} \right) = 0.$$

In other words, although div **J** is not zero, the divergence of the quantity $\mathbf{J} + \partial\mathbf{D}/\partial t$ is always zero, so that this quantity can mathematically be placed equal to a curl. Maxwell made the assumption that the last equation of (3.1) should properly be replaced by

$$\text{curl } \mathbf{H} = \mathbf{J} + \frac{\partial \mathbf{D}}{\partial t}. \tag{3.3}$$

The last term $\partial\mathbf{D}/\partial t$ is called the "displacement current," to distinguish it from **J**, the conduction current. By adding this term to Ampère's law, Maxwell assumed that a time rate of change of displacement produced a magnetic field, just as a conduction current does. A test of Maxwell's hypothesis can be made only with very rapidly varying fields, in which the rate of change of **D** with time is so great that the displacement current is large compared with the conduction current, or so that magnetic forces produced by it are comparable with the electric forces due to **E**. We shall see later that these cases are those in which we have electromagnetic waves, whose existence is possible only because of the presence of the displacement current in (3.3). Their existence, then, forms a demonstration of the correctness of Maxwell's hypothesis.

To understand the physical meaning of the displacement current in a simple case, consider the charging of a condenser. Current flows, from one plate through the wire to the other plate. If the current is i, this equals the rate of increase of charge on the plate. Suppose the plates are of area A, separation d, then the value of D between them is $D = \text{charge}/A$, and the displacement-current density in the region between plates is current$/A$. Thus the total displacement current is equal to the conduction current in the wire, so that the current is continuous through the circuit.

4. Maxwell's Equations.—We now can write Maxwell's equations,

$$\operatorname{curl} \mathbf{E} = -\frac{\partial \mathbf{B}}{\partial t}, \qquad \operatorname{div} \mathbf{B} = 0,$$

$$\operatorname{curl} \mathbf{H} = \mathbf{J} + \frac{\partial \mathbf{D}}{\partial t}, \qquad \operatorname{div} \mathbf{D} = \rho, \tag{4.1}$$

in which we note that the divergence equations follow from the curl equations by taking the divergence, using the equation of continuity (3.2), and integrating with respect to time. These are to be supplemented by the relations (2.1) of Chap. IV, and (2.4) of Chap. VI,

$$\mathbf{D} = \epsilon \mathbf{E}, \qquad \mathbf{B} = \mu \mathbf{H}. \tag{4.2}$$

These are often called the "constitutive equations." If the current density $\mathbf{J}$ obeys Ohm's law, we often include it also in the statement of the constitutive equations. This law is easily stated in differential form by considering a small volume, having length L in the direction of the current, and cross-sectional area A normal to the current. We apply Ohm's law in the form potential difference $= iR$. Here the potential difference is the field $\mathbf{E}$ times the length L of the volume, the current is the area times the current density $\mathbf{J}$, and the resistance is the specific resistance times L/A. Hence we have

$$EL = A\mathbf{J}\,\frac{L}{A} \times \text{specific resistance},$$

$$\mathbf{J} = \sigma \mathbf{E}, \tag{4.3}$$

where σ, the specific conductivity, is the reciprocal of the specific resistance. The unit of conductivity is the same as that of J/E; that is, its units are 1/(ohm-meter) = mhos per meter, where 1 mho is defined as the reciprocal of 1 ohm. It should be noted that 1 mho/m is $\frac{1}{100}$ of 1 mho/cm, the usual unit of conductivity. Ohm's law in the form (4.3) can be added to the relations (4.2).

Maxwell's equations, taken together with the constitutive equations, determine the field, when we are given the charges and currents. To make a complete set of dynamical principles, however, we need two more relations. First is the formula (1.1) of Chap. I,

$$\mathbf{F} = q(\mathbf{E} + \mathbf{v} \times \mathbf{B}),$$

giving the force acting on a charge moving with a given speed, or the corresponding force acting on the charge and current in unit volume,

$$\mathbf{F} = \rho\mathbf{E} + (\mathbf{J} \times \mathbf{B}). \tag{4.4}$$

Secondly, we must have a law determining the motion of charge in terms of force acting. If the charge is in a metallic conductor obeying Ohm's law, (4.3), that law provides the necessary relation. For electrons and ions moving in empty space, however, as in a vacuum or discharge tube, we must use Newton's law, that the force equals the mass times the acceleration, or the time rate of change of the momentum. With such a law, we find the field from the charge, the force from the field, and the motion from the force, obtaining therefore a complete system of dynamics.

Maxwell's equations, the constitutive equations, and the force equation, as just written, form the foundation of the whole of electromagnetic theory, and as far as is known are exactly correct, aside from the corrections resulting from the quantum theory, which in a sense change the whole formulation of the theory. They allow the derivation of the electromagnetic theory of light, and of electromagnetic waves in general, which we shall shortly take up. They hold even for particles moving nearly with the velocity of light, for which the theory of relativity must be used in discussing the mechanical part of the motion, but for which Maxwell's methods are still correct for finding the forces. In the older developments of electrical engineering, the so-called "lumped-circuit theory," it was possible to operate with the limiting case of slow variation with time, in which the displacement current could be neglected. The newer developments of microwaves and distributed constants, however, operate with the theory of electromagnetic waves, for which the displacement current is essential, and Maxwell's equations as formulated in this section, rather than the quasistationary form of them as given in (3.1), must be used.

5. The Vector and Scalar Potentials.—We observe that, if $\mathbf{B}$ depends on time, curl $\mathbf{E} \neq 0$, so that there is no potential for $\mathbf{E}$. The ordinary electrical potential is thus confined to static problems. Further, if $\mathbf{J}$ or $\partial\mathbf{D}/\partial t \neq 0$, there is no potential for $\mathbf{B}$. We have seen in Chap. V how a potential can be introduced for **B**: one uses a vector potential **A,** possible because div **B** = 0.   That is, we let

$$\mathbf{B} = \text{curl } \mathbf{A}. \tag{5.1}$$

We can do this even in the general case.   And it proves that we can use a scalar potential φ, reducing to the electrostatic potential in the case of a steady state, but different in other cases, by a special device.   The relation that proves to be satisfied is that

$$\mathbf{E} = -\text{ grad } \varphi - \frac{\partial \mathbf{A}}{\partial t}, \tag{5.2}$$

reducing to the familiar $\mathbf{E} = -\text{ grad } \varphi$ when the vector potential is independent of time.   To verify these statements, we substitute the expressions for **E** and **B** in Maxwell's equations, and see if they can be satisfied by proper choice of **A** and φ.   First we note that

$$\text{div } \mathbf{B} = \text{div curl } \mathbf{A}$$

is automatically zero, since the divergence of any curl is zero.   Similarly, remembering that the curl of any gradient is zero, we find that the equation curl $\mathbf{E} = -\ \partial \mathbf{B}/\partial t$ is automatically satisfied.

Next we must consider the other two equations, in **D** and **H**.   To find these quantities in terms of **E** and **B,** we must use the constitutive equations.   The relations are simple only if ϵ and μ are constants independent of position, and we consider only that case.   From the equation curl $\mathbf{H} - \partial \mathbf{D}/\partial t = \mathbf{J}$, we have, using the relation of vector analysis that

$$\text{curl curl } \mathbf{A} = \text{grad div } \mathbf{A} - \nabla^2 \mathbf{A},$$

the result

$$\nabla^2 \mathbf{A} - \epsilon\mu \frac{\partial^2 \mathbf{A}}{\partial t^2} - \text{grad}\left(\text{div } \mathbf{A} + \epsilon\mu \frac{\partial\varphi}{\partial t}\right) = -\mu \mathbf{J}.$$

Similarly, from the equation div $\mathbf{D} = \rho$, with a little manipulation, we find

$$\nabla^2 \varphi - \epsilon\mu \frac{\partial^2 \varphi}{\partial t^2} + \frac{\partial}{\partial t}\left(\text{div } \mathbf{A} + \epsilon\mu \frac{\partial\varphi}{\partial t}\right) = -\frac{\rho}{\epsilon}.$$

Now let us choose **A** and φ subject to the condition that

$$\text{div } \mathbf{A} + \epsilon\mu \frac{\partial\varphi}{\partial t} = 0. \tag{5.3}$$

Since div **A** is so far arbitrary, we can do this.   Then the equations for the potentials become

$$\nabla^2 \mathbf{A} - \epsilon\mu \frac{\partial^2 \mathbf{A}}{\partial t^2} = -\mu \mathbf{J},$$

$$\nabla^2 \varphi - \epsilon\mu \frac{\partial^2 \varphi}{\partial t^2} = -\frac{\rho}{\epsilon}. \tag{5.4}$$

If $\mathbf{A}$ and φ satisfy these equations, then, as we stated before, the fields determined from them by (5.1) and (5.2) satisfy Maxwell's equations. The equations for the potentials are of the form called "d'Alembert's equation," and as can be seen are extensions of Poisson's equation, obtained by adding the time derivatives. We observe that, in regions where there is no charge and current density, the potentials satisfy the wave equation, which is the homogeneous equation obtained by setting the right side of d'Alembert's equation equal to zero. We shall show in the next chapter that this means that φ and $\mathbf{A}$ are given by functions representing waves traveling with the velocity $1/\sqrt{\epsilon\mu}$, and that the same thing is true of the fields $\mathbf{E}$ and $\mathbf{H}$.

In (5.4) we have set up equations for $\mathbf{A}$ and φ in terms of prescribed values of the charge and current density. Sometimes, however, we wish to assume that the current density obeys Ohm's law, and that the charge density is zero, as we should have in the interior of a conductor. In that case, assuming (4.3) for $\mathbf{J}$, and proceeding in a similar manner, we find that in place of (5.3) we should assume

$$\mathrm{div}\, \mathbf{A} + \sigma\mu\varphi + \epsilon\mu \frac{\partial \varphi}{\partial t} = 0,$$

and in place of the equations (5.4) we have

$$\nabla^2 \mathbf{A} - \sigma\mu \frac{\partial \mathbf{A}}{\partial t} - \epsilon\mu \frac{\partial^2 \mathbf{A}}{\partial t^2} = 0,$$

$$\nabla^2 \varphi - \sigma\mu \frac{\partial \varphi}{\partial t} - \epsilon\mu \frac{\partial^2 \varphi}{\partial t^2} = 0. \tag{5.5}$$

We shall find that these equations, involving first as well as second time derivatives, represent damped or attenuated waves, as we should expect in an absorbing medium, such as a metallic conductor.

Problems

1. Two parallel conducting strips, of width w, distance of separation d (where $d \ll w$), indefinitely long, carry current in opposite directions, so as to form the two conductors of an electric circuit. Find the self-inductance of such a circuit per unit length.

2. Two concentric thin-walled hollow conducting cylinders, of radii r_1 and r_2, carry current in opposite directions. Find the self-inductance per unit length. Find how the result is changed if the inner conductor is a solid conducting rod, carrying current uniformly distributed through its interior.

3. Find the self-inductance of a circuit consisting of a hollow cylindrical conductor of radius r, parallel to an infinite conducting sheet that forms the return path of the current. (*Suggestion:* Use the method of images.)

4. A magnetic field points along the z axis, and its magnitude is proportional to time, and independent of position. Find the vector potential. Assuming that the scalar potential is zero, find the induced electric field. Prove by direct integration, using a circular path, that the law of induction holds.

5. Describe the magnetic field between the plates of a condenser while it is being charged up.

6. Starting from the induction law, show that the line integral of $\left(\mathbf{E} + \dfrac{\partial \mathbf{A}}{\partial t} \right)$ around a closed path is zero, where $\mathbf{A}$ is the vector potential. From this show that the curl of the above vector vanishes, and hence that $\mathbf{E} = -\operatorname{grad} \varphi - \dfrac{\partial \mathbf{A}}{\partial t}$, where φ is the scalar potential.

7. Derive the differential equations satisfied by $\mathbf{A}$ and φ for quasistationary processes.

8. Starting with the equation of continuity, and assuming Ohm's law, show that the charge density in a conductor obeys the equation $\dfrac{\sigma}{\epsilon} \rho + \dfrac{\partial \rho}{\partial t} = 0$. Show that any existing charge distribution within a conductor will be damped off exponentially, and find the time required for it to be reduced to $1/e$ of its initial value. Insert numerical values for copper, and find the value of this time, the relaxation time. Similarly find the relaxation time for a good insulator, such as quartz.

9. A long solenoid of n turns per unit length carries an alternating current of angular frequency ω and peak value I. If a nonmagnetic rod of conductivity σ just filling the solenoid is placed inside it, find the current distribution, the magnetic field, and the heating loss per unit length inside the copper rod. (The heat loss per unit volume per second equals $\mathbf{J} \cdot \mathbf{E}$.) Discuss the behavior of your solution for the limiting cases of large and small conductivities. How would your results be altered if the rod were coaxial with the solenoid but of smaller radius?

10. The outer conductor of a coaxial line as in Prob. 2 is a thin-walled cylinder of radius a and thickness t. Show that the contribution to the inductance per unit length of the line arising from the magnetic field in this conductor is given very nearly by

$$\frac{\mu_0}{6\pi} \frac{t}{a}, \qquad \text{where } \frac{t}{a} \ll 1.$$

11. Show that, in a nondissipative medium if the divergence of the vector potential is set equal to zero, instead of satisfying Eq. (5.3), the scalar potential satisfies Poisson's equation with ρ, the density of electric charge, a function of coordinates and time. Find the differential equation satisfied by the vector potential under these conditions.

CHAPTER VIII

ELECTROMAGNETIC WAVES AND ENERGY FLOW

The first and most conspicuous success of Maxwell's theory was its prediction of the existence of electromagnetic waves, whose velocity of propagation equaled the experimentally known velocity of light. It was immediately clear that light must be a form of electromagnetic radiation, of short wave length. It was a number of years later that Hertz demonstrated experimentally the existence of electromagnetic waves of longer wave length, but when they were found, they proved, like light, to satisfy Maxwell's equations. In the present chapter we shall discuss the simplest forms of waves, plane waves; we come to more complicated waves, such as spherical waves, in a later chapter. We also consider an aspect of the electromagnetic field that we have so far passed over: the energy associated with it. We have so far treated problems by considering only the forces acting, rather than the work done and the resulting energy. With radiation, however, it is so obvious that the field carries energy that without a consideration of the energy flow we can hardly form a correct picture of the phenomena. We therefore take up, not only the density of electric and magnetic energy in the field, which ties in with well-known facts regarding the energy in condensers and inductances, but also Poynting's theorem, a deduction from Maxwell's equations which provides a simple way of finding the flow of energy in any electromagnetic field.

1. Plane Waves and Maxwell's Equations.—Let us consider the problem of solving Maxwell's equations in a uniform material having dielectric constant $\kappa_e = \epsilon/\epsilon_0$, magnetic permeability $\kappa_m = \mu/\mu_0$, conductivity σ, but not any charge, or any current other than that determined by Ohm's law. Maxwell's equations in this case are

$$\text{curl } \mathbf{E} = -\mu \frac{\partial \mathbf{H}}{\partial t}, \qquad \text{div } \mathbf{H} = 0,$$

$$\text{curl } \mathbf{H} = \sigma \mathbf{E} + \epsilon \frac{\partial \mathbf{E}}{\partial t}, \qquad \text{div } \mathbf{E} = 0. \tag{1.1}$$

We take the curl of the first equation and substitute from the second for curl **H,** obtaining

$$\text{curl curl } \mathbf{E} = -\sigma\mu \frac{\partial \mathbf{E}}{\partial t} - \epsilon\mu \frac{\partial^2 \mathbf{E}}{\partial t^2}.$$

Similarly, we take the curl of the second and substitute from the first, obtaining

$$\text{curl curl } \mathbf{H} = -\sigma\mu \frac{\partial \mathbf{H}}{\partial t} - \epsilon\mu \frac{\partial^2 \mathbf{H}}{\partial t^2}.$$

Using the equation of vector analysis,

$$\text{curl curl } \mathbf{A} = \text{grad div } \mathbf{A} - \nabla^2\mathbf{A},$$

where **A** is an arbitrary vector function, and using the equations div $\mathbf{E} = 0$, div $\mathbf{H} = 0$, these equations become

$$\nabla^2\mathbf{E} - \sigma\mu \frac{\partial \mathbf{E}}{\partial t} - \epsilon\mu \frac{\partial^2 \mathbf{E}}{\partial t^2} = 0,$$

$$\nabla^2\mathbf{H} - \sigma\mu \frac{\partial \mathbf{H}}{\partial t} - \epsilon\mu \frac{\partial^2 \mathbf{H}}{\partial t^2} = 0. \tag{1.2}$$

Thus **E** and **H** satisfy the same wave equation that we have already found in Eq. (5.5), Chap. VII, for the potentials **A** and φ. We can of course get the case of a nonconducting medium by setting $\sigma = 0$, in which case the middle term of the equation is omitted. The equations (1.2) are vector equations, which means that each of the six components of **E** and **H** separately satisfies the same scalar wave equation.

With a wave equation of the form of

$$\nabla^2 u - \sigma\mu \frac{\partial u}{\partial t} - \epsilon\mu \frac{\partial^2 u}{\partial t^2} = 0, \tag{1.3}$$

where u, a scalar, can stand for one of the components of **E, H, A,** or φ, we can find a great variety of solutions. In fact, a good deal of the remainder of our study will be devoted to different solutions of this equation. We shall start with the simplest, and in many ways the most important, of these solutions, the plane wave, understanding that it is a special case, rather than a general solution. We shall assume that u is a function of z, one of the space coordinates, and of t, only, being independent of x and y, or constant on planes of constant z. Furthermore, we shall assume that u varies with both z and t in an exponential manner. We shall write our assumptions in the form

$$u = u_0 e^{j\omega t - \gamma z}, \tag{1.4}$$

where $j = \sqrt{-1}$, and u_0, ω, and γ are constants. We shall describe the physical significance of these constants in a moment. Substituting (1.4) in the wave equation (1.3), we find

$$\gamma = \pm \sqrt{\sigma\mu j\omega - \epsilon\mu\omega^2} = \pm j\omega \sqrt{\left(\epsilon - \frac{j\sigma}{\omega}\right)\mu}. \qquad (1.5)$$

In other words, if γ is given by either of the values (1.5), the expression (1.4) is a solution of the wave equation.

From the form of (1.5) we see that γ is in general complex. We may write it

$$\gamma = \alpha + j\beta, \qquad (1.6)$$

where α and β are real and imaginary parts, respectively. We note that the quantity whose square root is being taken in (1.5) lies in the second quadrant in the complex plane, its real part being negative and its imaginary part positive; thus the positive square root lies in the first quadrant, so that α and β are both positive. If in particular the medium is nonconducting, so that σ is zero, we have $\gamma = j\beta$, where β is positive. We may now rewrite the solution (1.4) in the form

$$u = u_0 e^{\mp \alpha z} e^{j(\omega t \mp \beta z)}. \qquad (1.7)$$

As in the theory of oscillating circuits, for physical purposes we use the real part of this complex expression to represent the real value of u. Thus (1.7) represents a disturbance whose value at a given point of space varies sinusoidally with time, with angular frequency ω, or frequency $f = \omega/2\pi$. It also varies sinusoidally with z, in case $\alpha = 0$, with wave length $\lambda = 2\pi/\beta$, so that, when z increases by a wave length, the disturbance reverts to its initial value. A given wave crest moves with a velocity v, given by

$$v = \frac{\omega}{\beta} = \lambda f,$$

the velocity being positive, or in the direction of increasing z, for the upper sign in (1.7), and negative for the lower sign. The factor $e^{\mp \alpha z}$ represents a falling off of intensity, or damping, in the direction of propagation of the wave, such that the amplitude falls to $1/e$ of its value in a distance $1/\alpha$. We see that the disturbance in a conducting medium is then a damped plane wave, and in a nonconducting dielectric it is an undamped plane wave.

For an undamped wave in a nonconducting medium, the velocity of propagation takes a simple form. In that case, setting $\sigma = 0$, we have from (1.5) and (1.6),

$$\beta = \omega \sqrt{\epsilon\mu}, \qquad v = \frac{\omega}{\beta} = \frac{1}{\sqrt{\epsilon\mu}}. \tag{1.8}$$

If we set

$$\epsilon = \kappa_e\epsilon_0, \qquad \mu = \kappa_m\mu_0,$$

following Eqs. (2.1) of Chap. IV, and (2.4) of Chap. VI, we find

$$v = \frac{c}{n}, \qquad c = \frac{1}{\sqrt{\epsilon_0\mu_0}} = 3.00 \times 10^8 \text{ m/sec},$$

$$n = \sqrt{\kappa_e\kappa_m}. \tag{1.9}$$

That is to say, the velocity of the electromagnetic waves in empty space is 3×10^8 m/sec, which is known to be the velocity of light. We have already mentioned that it was this fact which originally led to the conviction that Maxwell's equations furnished a theory of light, and that Hertz's discovery of electromagnetic waves has led to a series of waves of this type which can exist for frequencies over apparently an unlimited range. We now are familiar with waves ranging from ordinary radio waves at frequencies of a few kilocycles per second, short-wave radio at a few megacycles per second, microwaves with thousands or tens of thousands of megacycles per second, or wave lengths down to the order of a centimeter, through the infrared with wave lengths from fractions of a millimeter down to the visible spectrum with wave lengths of a few thousand angstroms, through the X-ray region with wave lengths down to a few hundredths of an angstrom, and from there through the gamma rays down to wave lengths whose lower limit is so far not known. All these waves satisfy the wave equation, all are essentially the same type, and in our subsequent treatment we shall consider them all.

The index of refraction n, defined in (1.9) as the ratio of the velocity of light in free space to the velocity in a material medium, is given in terms of the dielectric constant and magnetic permeability. Experimentally, this relation holds very accurately for radio waves, and in some materials for the infrared region, but it is almost never correct in the visible region or for shorter wave lengths. The reason will be taken up in a later chapter. It is essentially the fact that the dielectric constant is a function of frequency, rather than a true constant. The charges inside an atom or molecule, which become polarized to lead to the dielectric phenomena, as we have described in Chap. IV, have a certain mass and consequent inertia. For this reason they do not polarize instantaneously, but lag to some extent behind the field that polarizes them. The treatment of Chap. IV holds only for the steady state, or for the static dielectric constant. In a later chapter we shall take up the time variation, showing that we obtain a correct picture of the variation of the index of refraction with frequency, or of the phenomenon of dispersion.

2. The Relation between E and H in a Plane Wave.—If we substitute expressions of the form (1.4), with constant vectors $\mathbf{E}_0$ and $\mathbf{H}_0$ rather than the scalar u_0 multiplying the exponential, into Maxwell's equations (1.1), we find the following equations from the various components of Maxwell's equations:

$$\gamma E_y = -\mu j\omega H_x$$
$$-\gamma E_x = -\mu j\omega H_y$$
$$0 = -\mu j\omega H_z$$
$$-\gamma H_z = 0$$
$$\gamma H_y = (\sigma + \epsilon j\omega)E_x$$
$$-\gamma H_x = (\sigma + \epsilon j\omega)E_y$$
$$0 = (\sigma + \epsilon j\omega)E_z$$
$$-\gamma E_z = 0.$$

We notice in the first place that E_z and H_z are zero; that is, there is no component of $\mathbf{E}$ or $\mathbf{H}$ in the direction of propagation, or the wave is transverse. The remaining equations can be written in the form

$$\frac{E_x}{H_y} = -\frac{E_y}{H_x} = Z_0 = \frac{\mu j\omega}{\gamma} = \frac{\gamma}{\sigma + \epsilon j\omega} = \pm \sqrt{\frac{\mu}{\epsilon - j\sigma/\omega}}, \quad (2.1)$$

in which the various forms of the constant Z_0 are related through (1.5). These express the fact that $\mathbf{E}$ and $\mathbf{H}$ are at right angles to each other, as well as being at right angles to the direction of propagation, and that the ratio of the magnitude of $\mathbf{E}$ to the magnitude of $\mathbf{H}$ is the constant Z_0. We may write this expression in a vector form, if $\mathbf{k}$ is unit vector along the z axis; we have

$$\mathbf{k} \times \mathbf{E} = Z_0\mathbf{H}, \qquad \mathbf{k} \times \mathbf{H} = -\frac{\mathbf{E}}{Z_0}. \quad (2.2)$$

In the case of empty space, we have

$$Z_0 = \pm \sqrt{\frac{\mu_0}{\epsilon_0}} = \pm \sqrt{\frac{4\pi \times 10^{-7}}{8.85 \times 10^{-12}}} = \pm 376.6 \text{ ohms}, \quad (2.3)$$

where the units of Z_0 are most easily seen from the fact that it measures a ratio of $\mathbf{E}$, in volts per meter, to $\mathbf{H}$, in ampere-turns per meter, and therefore must equal volts/amperes or ohms. Because the units of

E/H are the same as those of impedance, the value of Z_0 is often referred to as the "wave impedance" of the medium, and the particular value (2.3) is the wave impedance of empty space, which in that particular case is a pure resistance, though we see from (2.1) that in a conducting medium there is a reactive as well as a resistive component.

3. Electric and Magnetic Energy Density.—In our treatment of electrostatics and magnetostatics we worked entirely with forces rather than introducing the concept of energy. In Sec. 3, Chap. I, however, we introduced the electrostatic potential, from which we can see at once that the energy of two charges q_1 and q_2, at a distance r_{12} from each other, is

$$V_{12} = \frac{q_1 q_2}{4\pi\epsilon_0 r_{12}}. \tag{3.1}$$

The total electrostatic energy of a system of charges is then the sum of expressions like (3.1) over all pairs of charges. A sum over pairs of terms like V_{ij} can also be written as half the double sum of the same quantities over the indices i and j separately (omitting of course the case $i = j$); the double sum includes each pair of indices twice, as for instance the terms V_{12} (for $i = 1, j = 2$) and V_{21}, which equals it (for $i = 2, j = 1$), and to compensate this we must have a factor $\frac{1}{2}$. Thus we have for the total potential energy

$$V = \frac{1}{2} \sum_i q_i \sum_{j \neq i} \frac{q_j}{4\pi\epsilon_0 r_{ij}}. \tag{3.2}$$

But the summation over j, in the expression above, is just the potential φ_i at the location of the charge q_i. Thus we may rewrite (3.2) in the form

$$V = \frac{1}{2} \sum_i q_i \varphi_i.$$

If the charges are continuously distributed, we may replace the summation over charges by an integration, replacing q_i by $\rho\, dv$, the charge in the volume dv. Thus we have

$$V = \frac{1}{2} \int \rho\varphi\, dv. \tag{3.3}$$

This is the standard expression for the energy of a system of charges, in terms of the charge density and the potential.

The expression (3.3) can be rewritten in terms of the field, rather than of the charge and potential. We have, for the electrostatic case,

$$\mathbf{E} = -\operatorname{grad} \varphi, \qquad \operatorname{div} \mathbf{D} = -\operatorname{div} (\epsilon \operatorname{grad} \varphi) = \rho,$$

which is the form that Poisson's equation would take in general where ϵ may be a function of position. Using this result, (3.3) takes the form

$$V = -\tfrac{1}{2}\int \varphi \ \text{div} \ (\epsilon \ \text{grad} \ \varphi) \ dv. \tag{3.4}$$

But by simple vector analysis we have

$$\text{div} \ (\epsilon \varphi \ \text{grad} \ \varphi) = \varphi \ \text{div} \ (\epsilon \ \text{grad} \ \varphi) + \epsilon(\text{grad} \ \varphi)^2.$$

Substituting in (3.4), the energy V becomes

$$V = \tfrac{1}{2}\int \epsilon E^2 \ dv + \tfrac{1}{2}\int \epsilon \varphi \mathbf{E} \cdot \mathbf{n} \ da.$$

If we integrate over such a large volume that φ and $\mathbf{E}$ are negligible around the surface, the surface integral vanishes, leaving us with

$$V = \tfrac{1}{2}\int \epsilon E^2 \ dv. \tag{3.5}$$

This is the expression for the electrostatic energy in terms of the field. Its interpretation is that we may imagine the energy to be localized throughout the field.

For instance, suppose we take a condenser of capacity C, and let its charge at a given moment be q. Assume that we are charging the condenser, and that we wish to know how much work we shall have to do on it to charge it. To take a small additional charge dq around the circuit, against the difference of potential q/C, will require an amount of work $(q/C) \ dq$. Thus the whole work done in setting up a charge Q is

$$\int_0^Q \frac{q}{C} \ dq = \frac{1}{2}\frac{Q^2}{C}.$$

But if the condenser consists of two plates of area A, distance of separation d, filled with a dielectric of dielectric constant ϵ/ϵ_0, the value of D inside the condenser is Q/A, and the value of E is $Q/\epsilon A$. Since the capacity C is $\epsilon A/d$, we then have

$$\frac{1}{2}\frac{Q^2}{C} = \frac{1}{2}\frac{(\epsilon A E)^2}{\epsilon A} d = \frac{1}{2}(\epsilon E^2)Ad.$$

But, since Ad is the volume of the region in which the field is different from zero, this agrees with the expression (3.5). Thus we see in this example how we can interpret (3.5) as meaning that energy is located throughout the field, with a density $\tfrac{1}{2}\epsilon E^2$.

In a similar way we can consider the magnetic energy to be localized in space. A complete treatment of the magnetic case is considerably more complicated than in the electrical case, and we shall give only a partial discussion; the reader who is interested in a more complete treatment may find one in J. A. Stratton, *Electromagnetic Theory*, Secs. 2.14 to 2.18 (McGraw-Hill Book Company, Inc., New York, 1941). In the first place, let us consider a circuit of self-inductance L, carrying an instantaneous current i. If i increases, the emf is $-L\, di/dt$, and the work that an external emf must do to bring about an increase di of the current is $Li(di/dt)\, dt = Li\, di$. Thus the total work that must be done to build up a current I in the circuit is

$$T = \int_0^I Li\, di = \tfrac{1}{2}LI^2.$$

Using Eqs. (2.1) and (2.2) of Chap. VII, this may be written

$$T = \tfrac{1}{2}\!\int\! I\, \mathbf{ds} \cdot \mathbf{A}. \tag{3.6}$$

In this expression, the integration is around the circuit, I is the current flowing in the element $\mathbf{ds}$, and $\mathbf{A}$ is the vector potential at that point. Using an argument like that above, we can also include the mutual effect when many currents are flowing; the only difference is that $\mathbf{A}$ in (3.6) must be the resultant vector potential not only of the one circuit but of all circuits. If the current densities are distributed throughout space, the current density being $\mathbf{J}$, we replace an integral about a circuit by a volume integral, and have

$$T = \tfrac{1}{2}\!\int\! \mathbf{J} \cdot \mathbf{A}\, dv. \tag{3.7}$$

The analogy of this expression to (3.3) is obvious, and it shows us for the first time that the vector potential has much the same relation to the energy of currents as the scalar potential has to the energy of charges.

We may now carry out a transformation, as with the electrostatic case, so as to write this energy T as an integration of the magnetic field rather than of the current density and vector potential. We have

$$\text{div } (\mathbf{H} \times \mathbf{A}) = \mathbf{A} \cdot \text{curl } \mathbf{H} - \mathbf{H} \cdot \text{curl } \mathbf{A} = \mathbf{J} \cdot \mathbf{A} - \mathbf{H} \cdot \mathbf{B},$$

in which we have used Ampère's law $\mathbf{J} = \text{curl } \mathbf{H}$ in the form that holds for the stationary case. Using this result we may rewrite (3.7) in a form involving a volume integral of $\mathbf{H} \cdot \mathbf{B}$, and a surface integral of the normal component of $\mathbf{H} \times \mathbf{A}$. As before, if we integrate over all space, we may assume that the fields are so small over the infinitely distant outer boundary that the surface integral vanishes, and we have finally

$$T = \tfrac{1}{2}\!\int\! \mathbf{H} \cdot \mathbf{B}\, dv = \tfrac{1}{2}\!\int\! \mu H^2\, dv, \tag{3.8}$$

in which the second form holds when $\mathbf{B} = \mu\mathbf{H}$. In case this is not so, and we have a ferromagnetic medium, the situation is much more complicated, and we shall not consider that case. The expression (3.8) is clearly analogous to (3.5) for the electrostatic energy. In the special case of the solenoid we may at once verify the result (3.8), as we did in the condenser for the electrostatic case. Let us take a solenoid of N turns, length d, area A, in a medium of permeability μ/μ_0. If a current i flows in it, the magnetic field inside will be $(N/d)i$, and the induction B will be $(N/d)\mu i$. The flux of B through the N turns will then be NA times B, or $(\mu N^2 A/d)i$, so that the self-inductance will be $L = \mu N^2 A/d$. The magnetic energy T is then

$$\frac{1}{2} Li^2 = \frac{1}{2} \frac{\mu N^2 A}{d} \left(\frac{Hd}{N}\right)^2 = \frac{1}{2} (\mu H^2) Ad,$$

which agrees with (3.8), and fits in with the assumption that the density of magnetic energy is $\frac{1}{2}\mu H^2$.

The expression (3.5) for the electrostatic energy is convenient for solving certain types of problems, because we can find the force on a charged body by stating that the work done when it is displaced is the change in the potential energy V, assuming the charges to remain fixed. For example, we can find the force acting to pull a piece of dielectric from one part of a field to another, by solving the problem of the field distribution as a function of the location of the piece of dielectric, computing the energy as a function of position, and seeing how it changes when the dielectric moves. The corresponding magnetic problem is much more involved, however, and is discussed by Stratton, in the reference given above.

If we take a circuit carrying a certain current, and displace it in a magnetic field, keeping the field and the current constant, the work that we must do proves not to be dT, as we should expect at first sight, but $-dT$. On the other hand, in moving the circuit through the field, the flux through it has changed, and this has resulted in an emf in the circuit. This emf would have resulted in a change of the current through the circuit, which is contrary to our hypothesis that the current is unchanged. Thus we must have been exerting a counter-emf to keep the current constant, and we find that the work that our counter-emf has had to do on the current during the displacement is just dT. Thus the work that our counter-emf does just balances the work that we get out of the system by the displacement of the circuit, and the net work done is zero. This is consistent with an observation that we might have made from Eq. (1.1), Chap. I, in which we state that the force exerted on a charge q moving with a velocity **v**, in a magnetic induction **B**, is $q\mathbf{v} \times \mathbf{B}$. This force is at right angles to the velocity of the charge, and therefore the work done by it is zero.

On the other hand, we are assuming that the field stays constant while the circuit is being moved. Actually this field must be produced by other currents in other circuits. As the circuit is moved, the flux produced by it at these other circuits changes, and emf's are induced in them. To keep the field constant, and hence the currents in those circuits constant, we must have been exerting counter-emf's in those other circuits as well, and the work we have done to maintain those counter-emf's proves again to be dT. Thus in the process we have a net amount of work dT, made up of $-dT$ as the direct work involved in moving the circuit, dT in keeping the current constant in the circuit we are moving, and another dT in keeping the current constant in the circuits that produce the external magnetic field. Unless careful account is taken of these various terms in the energy, whose existence we have merely mentioned without proof, there is great danger of making a mistake in sign when finding the mechanical forces acting on a circuit from the magnetic energy.

These relations are similar to those found in mechanics in certain cases, in which the kinetic energy depends on the coordinates. In such a case we find that the force is given by the derivative, not of the energy, but of the difference between the potential and kinetic energies, with respect to the coordinates. This difference is essentially the Lagrangian function. In a similar way here the force is given by the derivative, not of the energy $V + T$, but of the difference $V - T$, which plays the part of a Lagrangian function, the magnetic energy T playing the part of a kinetic energy. This is not unnatural; for the electrostatic energy V results from the positions of charges, but the magnetic energy T results from their motions, the current i in the expression $\frac{1}{2}Li^2$ being proportional to the velocities of the charges.

4. Poynting's Theorem and Poynting's Vector.—In the preceding sections we have seen that in the static case we have an energy density $\frac{1}{2}(\epsilon E^2 + \mu H^2)$ of electric and magnetic energy. We shall now show that this formula still can be used when the fields are changing with time. To show this we shall first prove a mathematical theorem, Poynting's theorem, and shall then show its interpretation. By a vector identity we have

$$\text{div } (\mathbf{E} \times \mathbf{H}) = \mathbf{H} \cdot \text{curl } \mathbf{E} - \mathbf{E} \cdot \text{curl } \mathbf{H}$$

$$= -\mathbf{E} \cdot \frac{\partial \mathbf{D}}{\partial t} - \mathbf{H} \cdot \frac{\partial \mathbf{B}}{\partial t} - \mathbf{E} \cdot \mathbf{J},$$

where in the last expression we have used Maxwell's equations. If we assume $\mathbf{D} = \epsilon\mathbf{E}$, $\mathbf{B} = \mu\mathbf{H}$, we have

$$\text{div } (\mathbf{E} \times \mathbf{H}) + \frac{\partial}{\partial t}\frac{1}{2}(\epsilon E^2 + \mu H^2) = -\mathbf{E} \cdot \mathbf{J}. \qquad (4.1)$$

This equation reminds us of an ordinary equation of continuity, which states that the divergence of the flux of any quantity, plus the rate at which the density of the quantity increases with time, equals the rate at which the quantity is produced. In other words, applying the equation to a small volume, it states that the rate at which the quantity increases within the volume equals the rate at which it is produced within the volume, minus the rate at which it flows out over the surface. The quantity $-\mathbf{E} \cdot \mathbf{J}$ represents the rate at which energy is produced (that is, $\mathbf{E} \cdot \mathbf{J}$ represents the rate at which energy is lost) per unit volume on account of ordinary joulean or resistance heating. Thus the quantity for which (4.1) forms the equation of continuity is the energy. We are then justified in interpreting the quantity $\frac{1}{2}(\epsilon E^2 + \mu H^2)$ as the energy density, extending the results of the preceding sections to time varying fields. At the same time, we must interpret the vector

$$\mathbf{S} = \mathbf{E} \times \mathbf{H},$$

which is known as "Poynting's vector," as the flux of energy, the amount of energy crossing unit area perpendicular to the vector, per unit time.

The conception of the energy of the electromagnetic field as residing in the medium is a very fundamental one, and has had great influence in the development of the theory. Maxwell and his immediate followers thought of the medium as resembling an elastic solid, the electrical energy representing the potential energy of strain of the medium, the magnetic energy the kinetic energy of motion. Though such a mechanical view is no longer held, still the energy is regarded as being localized in space, and as traveling in the manner indicated by Poynting's vector. Thus, in a light wave, we shall show that there is a certain energy per unit volume, proportional to the square of the amplitude (E or H). This energy travels along, and Poynting's vector is the vector that measures the rate of flow, or the intensity of the wave. We have already seen that $\mathbf{E}$ and $\mathbf{H}$ in a plane wave are at right angles to each other, and at right angles to the direction of flow; thus $\mathbf{E} \times \mathbf{H}$ must be along the direction of flow. In more complicated waves as well, Poynting's vector points along the direction of the flow of radiation. If, for example, we have a source of light, and we wish to find at what rate it is emitting energy, we surround it by a closed surface, and integrate the normal component of Poynting's vector over the surface. The whole conception of energy being transported in the medium is fundamental to the electromagnetic theory of light.

When we come to charges and currents, however, it is a little harder to see the significance of the energy in the medium. For example, in a circuit consisting of a battery, and a wire connecting the plates, Poynting's vector indicates that the energy flows out of the battery through the space surrounding the wire, and finally flows into the wire at the point where it will be transformed into heat. This seems to have small physical significance. With a moving charge, the situation is somewhat more reasonable. A simple model of an electron, which was supposed before the quantum theory to represent its actual structure, was a sphere of radius R, on the surface of which the charge is distributed. Then the field E will be $e/4\pi\epsilon_0 r^2$ at any point outside the sphere, where e is the charge. The total electrical energy is the volume integral of $(e^2/32\pi^2\epsilon_0)(1/r^4)$ over all space outside the sphere, or

$$\frac{e^2}{32\pi^2\epsilon_0} \int_R^\infty \frac{4\pi r^2}{r^4}\, dr = \frac{e^2}{8\pi\epsilon_0 R}.$$

In the classical theory of the electron, which we have mentioned, it is this quantity which is interpreted as being the actual constitutive energy of the electron, though a correction must be made of an additional energy of a nonelectromagnetic nature that is required to keep the sphere in equilibrium. Neglecting this correction, we can compute the mass of the electron. For a relation of Einstein's relativity theory says that a given energy has a mass, given by the relation energy = mc^2. Hence $mc^2 = e^2/8\pi\epsilon_0 R$. Solving for the radius, we have $R = e^2/8\pi\epsilon_0 mc^2$, a familiar formula for the radius of the electron. The more correct formula, inserting the correction we omitted, differs only by a small factor. Using the values $e = 1.60 \times 10^{-19}$ coulomb, $m = 9.1 \times 10^{-31}$ kg, $c = 3.00 \times 10^8$ m/sec, $\epsilon_0 = 8.85 \times 10^{-12}$ farad/m, we have $R = 1.42 \times 10^{-15}$ m $= 1.42 \times 10^{-13}$ cm. Now, if this electron moves, it will produce a magnetic field, as a current would, and hence will have a certain magnetic energy. Since the magnetic field is proportional to the velocity (or the current), the magnetic energy is proportional to the square of the velocity. This can be shown to be the kinetic energy. Further, there will be a Poynting vector, pointing in general in the direction of travel of the electron, and representing the flow of energy associated with the electron. All these relations prove on closer examination to be more complicated than they seem at first sight, but they are suggestive in pointing one possible way to an eventual theory of the structure of the electron, which even the present quantum theory is unable to supply completely.

5. Power Flow and Sinusoidal Time Variation.—We shall often want to find Poynting's vector, and the energy density, in cases where the fields **E** and **H** are the real parts of complex exponentials such as that given in (1.7). At a given point of space, let us assume that **E** is given by the real part of $\mathbf{E}_0 e^{j\omega t}$, and **H** by the real part of $\mathbf{H}_0 e^{j\omega t}$, where $\mathbf{E}_0$ and $\mathbf{H}_0$ are complex vector functions of position. Let the real part of $\mathbf{E}_0$ be $\mathbf{E}_r$, and the imaginary part $\mathbf{E}_i$, with similar notation for $\mathbf{H}_0$. Then **E** is given by

$$Re(\mathbf{E}_0 e^{j\omega t}) = Re[(\mathbf{E}_r + j\mathbf{E}_i)(\cos \omega t + j \sin \omega t)] = \mathbf{E}_r \cos \omega t - \mathbf{E}_i \sin \omega t.$$

H is given by a similar expression. Poynting's vector is then

$$\mathbf{E} \times \mathbf{H} = (\mathbf{E}_r \times \mathbf{H}_r) \cos^2 \omega t + (\mathbf{E}_i \times \mathbf{H}_i) \sin^2 \omega t \\ - [(\mathbf{E}_r \times \mathbf{H}_i) + (\mathbf{E}_i \times \mathbf{H}_r)] \sin \omega t \cos \omega t.$$

We notice that there are two types of terms: the first two, whose time average is different from zero, since $\cos^2 \omega t$ and $\sin^2 \omega t$ average to $\frac{1}{2}$; and the last term, whose time average is zero, since $\sin \omega t \cos \omega t$ averages to zero. Thus the time average Poynting vector is

$$\text{Average } (\mathbf{E} \times \mathbf{H}) = \frac{1}{2}(\mathbf{E}_r \times \mathbf{H}_r + \mathbf{E}_i \times \mathbf{H}_i). \tag{5.1}$$

This can be rewritten in a convenient way, by using the notation of complex conjugates, where the complex conjugate of a complex number is the number obtained from the original one by changing the sign of j wherever it appears, and is indicated by a bar over the number. In terms of this notation, let us consider the quantity $(\mathbf{E} \times \overline{\mathbf{H}})$. This is

$$\begin{aligned}(\mathbf{E} \times \overline{\mathbf{H}}) &= (\mathbf{E}_0 e^{j\omega t}) \times (\overline{\mathbf{H}}_0 e^{-j\omega t}) \\ &= (\mathbf{E}_0 \times \overline{\mathbf{H}}_0) = (\mathbf{E}_r + j\mathbf{E}_i) \times (\mathbf{H}_r - j\mathbf{H}_i) \\ &= (\mathbf{E}_r \times \mathbf{H}_r + \mathbf{E}_i \times \mathbf{H}_i) + j(\mathbf{E}_i \times \mathbf{H}_r - \mathbf{E}_r \times \mathbf{H}_i). \end{aligned} \tag{5.2}$$

We see that, except for the factor $\frac{1}{2}$, the real part of (5.2) is just the same as the quantity appearing in (5.1). That is, we have

$$\text{Average } (\mathbf{E} \times \mathbf{H}) = \frac{1}{2} Re(\mathbf{E} \times \overline{\mathbf{H}}), \tag{5.3}$$

where the **E** and **H** appearing on the right side of the equation are the complex quantities whose real parts give the real **E** and **H** appearing on the left of the equation. A similar derivation for the energy density shows that

$$\text{Average } \frac{1}{2}\epsilon E^2 = \frac{1}{4}\epsilon \mathbf{E} \cdot \overline{\mathbf{E}}, \tag{5.4}$$

with a similar formula for the magnetic energy.

6. Power Flow and Energy Density in a Plane Wave.—In (2.2) we have found that **E** and **H** in a plane wave are at right angles to each other, and at right angles to the direction of propagation. In particular, we have for a nonabsorbing medium, where Z_0 is real,

$$\frac{1}{2} Re(\mathbf{E} \times \bar{\mathbf{H}}) = \frac{1}{2} \frac{\mathbf{k}}{Z_0} (\mathbf{E} \cdot \bar{\mathbf{E}}) = \frac{1}{2} \mathbf{k} Z_0 (\mathbf{H} \cdot \bar{\mathbf{H}}). \tag{6.1}$$

That is, the flow of energy is along the direction of propagation, and proportional to the square of the amplitude of **E** or of **H**. Similarly the average electrical energy density is $\frac{1}{4}\epsilon(\mathbf{E} \cdot \bar{\mathbf{E}})$, and the average magnetic density is $\frac{1}{4}\mu(\mathbf{H} \cdot \bar{\mathbf{H}})$. Because of the relation (2.1) between the amplitudes of **E** and **H**, we find easily that the average magnetic energy equals the average electrical energy, so that the total energy density is $\frac{1}{2}\epsilon(\mathbf{E} \cdot \bar{\mathbf{E}})$. We note, from (6.1) and (2.1), together with (1.8), that we have the relation

$$\text{Energy flux} = v \times \text{energy density.} \tag{6.2}$$

This equation has a simple meaning. If the energy were flowing with velocity v, in the direction of propagation of the wave, all the energy contained in a cylinder of unit cross section, and height equal to v, would cross unit cross section per second, forming the flux.

In a conducting medium, in which α is different from zero, **E** and **H** will each contain a factor $e^{-\alpha z}$, so that Poynting's vector and the energy density will each have a factor $e^{-2\alpha z}$, showing that the intensity is damped or attenuated by this factor in traveling along. Poynting's theorem of course tells us that the energy lost to the wave as it advances is used up in resistive heating of the medium. We readily find that, in a conducting medium, the electrical energy is no longer equal to the magnetic energy, but as the conductivity becomes greater and greater, the electric energy becomes smaller and smaller compared with the magnetic energy, finally vanishing in the limit of infinite conductivity. We shall investigate these relations more completely in the next chapter.

Problems

1. If the generation of heat per unit volume in a conductor carrying a current is σE^2, prove that, for a cylindrical conductor of resistance R, carrying a current i, the rate of generation is i^2R.

2. Given a cylindrical wire carrying a current. Find the values of E and H on the surface of the wire, computing Poynting's vector, and show that it represents a flow of energy into the wire. Show that the amount flowing into a given length of wire is just enough to supply the energy that appears as heat in the length. Note that the surface of a wire carrying current is not an equipotential, so that there can be a component of electric field parallel to it.

3. Calculate the electrical and magnetic energies in a plane wave traveling in a conductor, and show by direct comparison that they are different from each other. What happens in the limiting cases $\sigma \to 0$ and $\sigma \to \infty$, that is, insulators and perfect conductors?

4. In the absence of charges, solutions of Maxwell's equations may be obtained from a vector potential **A** alone, setting the scalar potential equal to zero. Work this out for the general case of plane waves propagated along the z axis, showing that the vector potential can have two arbitrary complex amplitudes for its x and y components, the z dependence being in the factor $e^{i\omega t - \gamma z}$. From this find the relations between **E** and **H** in a plane wave, showing that your results are identical with those obtained in the text.

5. A coaxial line of length l, inner and outer radii a and b, carries a steady current I. The resistance of the inner conductor is R_i, that of the outer conductor R_0, and the load resistance is R. Find expressions for the components of **E** and **H** at any point between the conductors. Show that the component of E parallel to the axis reverses direction as one moves from the inner to the outer conductor, and that it is zero at a radius r_0 given by $\ln (r_0/a) = [R_i/(R_0 + R_i)] \ln (b/a)$. Compute the Poynting vector, and discuss the power flow in this field. Integrate the radial component of the Poynting vector over the surface of both inner and outer conductor, and show that these integrals yield the I^2R losses in each. Apply the Poynting theorem to a volume bounded by the planes z_1 and z_2 and the cylindrical surfaces r_1 and r_2.

6. Consider a plane wave in empty space given by

$$E_x = \sqrt{(\mu_0/\epsilon_0)}\, H_y = E_0 e^{i(\omega t - \beta z)},$$

with $\beta = \omega \sqrt{\epsilon_0 \mu_0}$. Show that an approximate solution to Maxwell's equations can be obtained to represent a plane wave of finite cross section by considering E_0 to be a slowly varying function of x and y and adding to the above field components the longitudinal components

$$E_z = -\frac{j}{\beta}\frac{\partial E_0}{\partial x}\, e^{i(\omega t - \beta z)} \qquad \text{and} \qquad H_z = -\frac{j}{\beta}\sqrt{\frac{\epsilon_0}{\mu_0}}\frac{\partial E_0}{\partial y}\, e^{i(\omega t - \beta z)}.$$

By slowly varying is meant that the percentage change in the fields is small compared with unity if one moves a distance of one wave length in a direction transverse to the direction of propagation, so that second derivatives and products of first derivatives with respect to x and y may be neglected.

7. Prove that, if a piece of steel is magnetized by an alternating current, the energy loss per cycle due to hysteresis is given by $\oint \mathbf{H} \cdot d\mathbf{B}$ per unit volume.

8. A magnetic circuit of uniform cross section consists of two sections separated by air gaps of length so small that fringing may be neglected. Consider the process of increasing the air-gap lengths each by an infinitesimal amount, maintaining the magnetizing current constant. Compute the change in magnetic field energy and the work done by the emf that maintains the magnetizing current constant during the displacement. From this obtain the mechanical work done in effecting the displacement, and show that the force with which the two sections attract each other is given by $B^2 A/\mu_0$, where A is the cross section of the magnetic circuit.

CHAPTER IX

ELECTRON THEORY AND DISPERSION

In Chap. VIII we investigated the propagation of plane waves in a medium with a given dielectric constant and conductivity. We have seen that in a nonconducting nonmagnetic medium a wave is propagated with a velocity c/n, where c is the velocity of light in empty space, n the index of refraction, which is equal to $\sqrt{\kappa_e}$. In a conducting medium, there is a damped wave, whose propagation constant, determining both the velocity of propagation and the rate of absorption, is given by Eq. (1.5), Chap. VIII, predicting a definite variation of propagation properties with frequency. These two formulas give a very straightforward way of testing the electromagnetic theory of light: we can measure the dielectric constant and conductivity of a material, and see if its optical index of refraction and absorption coefficient are properly related to these constants. One of the first observations after the formulation of Maxwell's theory was that these relations are not fulfilled by real substances. The departures between observation and the simple theory have been very useful in gaining a knowledge of the structure of actual dielectrics and conductors. Briefly, the discrepancy between theory and experiment is explained by supposing that the dielectric constant and conductivity are functions of frequency; a large part of the electron theory of solids is devoted to an explanation of the nature of this frequency variation.

The experimental situation is much clearer than it was in Maxwell's day, because of the wider ranges of the spectrum that have been explored in the meantime. We shall first describe the situation for dielectrics without conductivity. For long radio wave lengths, waves can be propagated with a velocity given by the index of refraction determined from the static dielectric constant. As the frequency increases, however, the index of refraction starts to increase, until somewhere in the spectrum it goes through a maximum, then drops suddenly, and begins to increase again. The phenomenon of change of index of refraction with frequency is called "dispersion," and the sudden drop we have just mentioned is called "anomalous dispersion." In the neighborhood of a region of anomalous dispersion, there is absorption, even when the material does not absorb elsewhere; the index of refraction and absorption coefficient in this neighborhood behave much like the reactance and resistance of a resonant electric circuit. Most materials have several regions of anomalous dispersion and absorption, in different parts of the spectrum. Many materials, for instance, have absorption in the infrared, and practically all materials absorb in the ultraviolet. Materials that absorb in the visible part of the spectrum are those which appear colored. By the time we go through the ultraviolet to the X-ray region, anomalous dispersion has stopped, and the index of refraction of all materials has become almost exactly unity; it is for this reason that lenses and prisms are practically impossible in the X-ray region. The absorption has also largely decreased in the X-ray region, and it is for that reason that X rays can penetrate so many types of materials. Passing beyond, to the region of gamma rays, all materials become rather transparent and nonrefracting.

For conductors, we found in Eq. (1.5), Chap. VIII, that γ is given by $j\omega \sqrt{(\epsilon - j\sigma/\omega)\mu}$. We see that as the frequency approaches zero the second term in the radical becomes infinitely greater in magnitude than the first. Thus at low frequencies the optical properties of a conductor are determined entirely by the conductivity, and not by the dielectric constant; in fact, there is no experimental way of finding the dielectric constant of a good conductor at low frequencies. The values of σ for good conductors, such as metals, are such that σ/ϵ does not become comparable with ω until we reach frequencies in the visible part of the spectrum. It is found experimentally that the optical properties of metals are well described by this simple theory through the radio-frequency, microwave, and infrared parts of the spectrum, and that, as we should expect, the dielectric effects begin to be important in the visible region. The situation there, however, is much more complicated than would be indicated by Eq. (1.5), Chap. VIII; for by then, there is good evidence that both ϵ and σ vary in complicated ways with the frequency. The variation of ϵ for a metal is not unlike that for a dielectric; there is anomalous dispersion, and consequent absorption, superposed on the conductivity. As we go through the ultraviolet to the X-ray region, the effect of conductivity becomes negligible, and metals are no more absorbing than other materials.

We shall now examine the way in which simple electronic models of dielectrics and conductors can lead to variations of dielectric constant and conductivity with frequency which are in at least qualitative agreement with the observations.

1. Dispersion in Gases.—In Chap. IV, we described the nature of dielectric polarization, stating that it can arise in two ways: from the polarization of the atoms or molecules, by the displacement of the electrons in them; and by orientation of dipoles already existing, under the action of the external field. We shall discuss only the first type of polarization in the present chapter. We shall find in a later section that, in a solid or liquid, the problem of polarization is complicated by the interactions between the molecules or atoms; but in a gas, the molecules are far enough apart so that we can neglect the interactions between them. Each molecule contains charges that can be displaced under the action of an external field, and these charges act as if they were held to positions of equilibrium by restoring forces proportional to the displacement. Thus in a static case an electron of charge e is acted on by the force eE of the external electric field, and $-ax$, a linear restoring force, proportional to the displacement x, with a constant of proportionality a. The displacement is then $x = (e/a)E$, and the induced dipole moment $ex = (e^2/a)E$. Thus the polarizability α of a molecule is e^2/a, and using Eq. (2.4) of Chap. IV, the dielectric constant is given by

$$\kappa_e = 1 + N \frac{e^2}{a\epsilon_0}, \tag{1.1}$$

where N is the number per unit volume.

The value (1.1) is the static value. If the external field varies with time, we must take account of the fact that the electron has a mass m, thus possessing inertia. We shall also introduce, in addition to the external electric force and the elastic restoring force, a damping force proportional to the velocity, to account for the absorption. The equation of motion for the electron is then

$$m \frac{d^2x}{dt^2} + mg \frac{dx}{dt} + m\omega_0^2 x = eE, \tag{1.2}$$

where we have rewritten the linear restoring force in terms of a constant $m\omega_0^2$. If E varies sinusoidally with time, as the real part of an expression containing the exponential factor $e^{j\omega t}$, we can then solve (1.2), as always in determining the forced oscillations of a linear oscillator, by assuming that x varies also as $e^{j\omega t}$. We then find

$$x = \frac{(e/m)E}{\omega_0^2 - \omega^2 + j\omega g}.$$

That is, the electron vibrates with the same frequency as the external field, but with an amplitude depending on the frequency. If we have

N_k electrons per unit volume characterized by constants ω_k and g_k, the polarization is

$$P = E \sum_k \frac{N_k e^2/m}{\omega_k^2 - \omega^2 + j\omega g_k},$$

and the dielectric constant is

$$\kappa_e = 1 + \sum_k \frac{N_k e^2/m\epsilon_0}{\omega_k^2 - \omega^2 + j\omega g_k}. \tag{1.3}$$

In the limit of low frequencies, where ω can be neglected, this reduces essentially to the value (1.1), the static dielectric constant, but at each of the resonant frequencies ω_k there is a phenomenon such as is found with the impedance of an electric circuit, near its resonant frequency. The magnitude of the dielectric constant varies rapidly with frequency, and at the same time it becomes complex, leading to absorption, as well as dispersion.

It is convenient, in an absorbing medium, to introduce an absorption coefficient k, as well as an index of refraction n, by the relation

$$\gamma = j\frac{\omega}{c}(n - jk), \qquad e^{-\gamma z} = e^{-(\omega/c)kz}e^{-j(\omega/c)nz}.$$

Then we have

$$(n - jk)^2 = \kappa_e.$$

For a gas, there are few enough molecules so that the second term of (1.3) is small compared with unity. Thus we have approximately

$$n - jk = 1 + \frac{1}{2}\sum_k \frac{N_k e^2/m\epsilon_0}{\omega_k^2 - \omega^2 + j\omega g_k},$$

and, if we separate into real and imaginary parts, we obtain

$$n = 1 + \frac{1}{2}\sum_k \frac{(N_k e^2/m\epsilon_0)(\omega_k^2 - \omega^2)}{(\omega_k^2 - \omega^2)^2 + \omega^2 g_k^2}$$

$$k = \frac{1}{2}\sum_k \frac{(N_k e^2/m\epsilon_0)\omega g_k}{(\omega_k^2 - \omega^2)^2 + \omega^2 g_k^2}.$$

If we consider these two quantities as functions of frequency, they show the properties we have already discussed. As the frequency increases, the index of refraction goes through a phenomenon of anomalous dispersion in the neighborhood of each of the resonance frequencies ω_k, the index eventually approaching unity, and being in fact slightly less than unity, for very large values of frequency; while the absorption coefficient is small everywhere except near each of the resonance frequencies. The behavior of n and k near resonance is as shown in Fig. 24.

2. Dispersion in Liquids and Solids.—In the case of solids and liquids we may no longer make the approximation that the force acting on an electron is simply the electric vector of the electromagnetic wave in free space, but must take into account the added force on the electron due to the polarization of the body. We can calculate this force as follows: we imagine a small sphere of radius R (with its center at the position of the electron in question) cut out of the medium. If we do this, without disturbing the distribution of polarization outside the sphere, we have induced charges on the surface

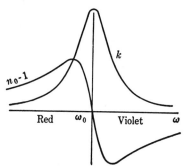

Fig. 24.—Anomalous dispersion, showing index of refraction and absorption coefficient as function of frequency.

of this spherical volume from which we calculate the force at the center of the sphere. The surface density of induced charge on a spherical ring at an angle θ to the direction of the field is $P \cos \theta$. Since the area of a ring included between angles θ and $\theta + d\theta$ is $2\pi R^2 \sin \theta \, d\theta$, the charge on the ring is $2\pi P R^2 \cos \theta \sin \theta \, d\theta$. This charge produces a field at the center of the sphere whose component parallel to E is

$$dE_1 = \frac{2\pi P R^2 \cos^2 \theta \sin \theta \, d\theta}{4\pi\epsilon_0 R^2},$$

so that the total charge on the spherical surface produces a field at the center equal to

$$E_1 = \frac{P}{2\epsilon_0} \int_0^\pi \cos^2 \theta \sin \theta \, d\theta = \frac{P}{3\epsilon_0}.$$

The total electric field at the center of this sphere is then

$$E + \frac{P}{3\epsilon_0}.$$

Of course, there is still the contribution to the force by the atoms inside the little sphere we have cut out, but in an isotropic medium this averages to zero.

We can now carry over our calculations for gases if we replace E by $(E + P/3\epsilon_0)$ in the expression for x. Thus we get

$$P = \left(E + \frac{P}{3\epsilon_0}\right) \sum_k \frac{N_k e^2/m}{\omega_k^2 - \omega^2 + j\omega g_k},$$

and using the relations $D = \kappa_e \epsilon_0 E = \epsilon_0 E + P$, we have

$$E + \frac{P}{3\epsilon_0} = \frac{\kappa_e + 2}{3} E,$$

and we find for κ_e

$$\frac{\kappa_e - 1}{\kappa_e + 2} = \frac{(n - jk)^2 - 1}{(n - jk)^2 + 2} = \frac{1}{3} \sum_k \frac{N_k e^2/m\epsilon_0}{\omega_k^2 - \omega^2 + j\omega g_k}.$$

If N represents the number of atoms per unit volume, and if f_k is the fraction of atoms of the kth type, so that $N_k = f_k N$, we may rewrite it

$$\frac{(n - jk)^2 - 1}{(n - jk)^2 + 2} \frac{1}{N} = \frac{1}{3} \sum_k \frac{f_k e^2/m\epsilon_0}{\omega_k^2 - \omega^2 + j\omega g_k}. \tag{2.1}$$

In this formula, the quantities f_k, which are independent of the number of atoms per unit volume, are often called the "oscillator strengths" of the various resonances.

For a transparent substance, where we can neglect the damping force, and the index of refraction is real, we have for a given frequency of light

$$\frac{n^2 - 1}{n^2 + 2} \frac{1}{\rho_0} = \text{constant},$$

where ρ_0 is the density of the body, obviously proportional to N. This law, known as the Lorenz-Lorentz law, is surprisingly well obeyed for many substances. In many cases it even gives approximately correctly the relationship between the index of refraction of a liquid and of its vapor. In the limit of very long electromagnetic waves, and for the electrostatic case, we have

$$\frac{\kappa_e - 1}{\kappa_e + 2} \frac{1}{\rho_0} = \text{constant},$$

the so-called "Clausius-Mosotti relation" between dielectric constant and density.

There is a different way of handling dispersion in liquids and solids, equivalent to this, but exhibiting the result in different form.

We shall consider only the case where there is but one type of oscillating electron. In (1.2) we replace the right side by $e(E + P/3\epsilon_0)$, but also we write $P = Nex$, which would be the relation for only one type of electron, and incorporate the term in x in the left side of the equation, so that the equation becomes

$$m \frac{d^2x}{dt^2} + mg \frac{dx}{dt} + m \left(\omega_0^2 - \frac{Ne^2}{3\epsilon_0 m} \right) x = eE.$$

Solving as in (1.3), we then find

$$(n - jk)^2 = 1 + \frac{Ne^2/m_0}{\bar{\omega}_0^2 - \omega^2 + j\omega g}, \tag{2.2}$$

where

$$\bar{\omega}_0^2 = \omega_0^2 - \frac{Ne^2}{3\epsilon_0 m}. \tag{2.3}$$

That is, we have the same type of anomalous dispersion that we found in (1.3) for a gas, but with the resonant frequency displaced according to (2.3). If there are various resonant frequencies, it is no longer a simple matter to prove that a solution like (1.3) can be set up, but the proof can be given, by methods similar to those used in discussing the normal coordinates of a coupled vibrating system in classical mechanics. We note that, since N is proportional to the density, the resonant frequencies appearing in a formula of the type of (2.2) vary with the density, whereas the frequencies appearing in the Lorenz-Lorentz formula (2.1) are independent of density.

3. Dispersion in Metals.—The simple electron theory of metals assumes that free electrons wander about among fixed ions, and carry the current. On the average there is no resultant force acting on the electrons, except the external field, so that there is a force eE acting on them. The equation of motion is thus similar to (1.2), except that there is no restoring force, so that ω_0 is zero. We must assume a damping force, proportional to the velocity, and we find that it is simply related to the conductivity, according to Ohm's law. For in the case of a constant external field, the electrons will acquire constant velocity, which by (1.2) will be eE/mg, so that, if N is the number of conducting electrons per unit volume, the current density is

$$J = Nev = \frac{Ne^2E}{mg} = \sigma E, \qquad \sigma = \frac{Ne^2}{mg}. \tag{3.1}$$

This holds only for a steady field, however. If E varies as $e^{i\omega t}$, then we find from (1.2) that

$$\sigma = \frac{Ne^2}{mg + mj\omega},$$

so that the conductivity varies with frequency, vanishing with infinite frequency. In considering propagation, we have seen in Eq. (1.5), Chap. VIII, that the combination $\kappa_e - j\sigma/\omega\epsilon_0$ is to be used in place of the dielectric constant in a transparent medium. Using (1.3) for κ_e, this gives

$$\kappa_e - \frac{j\sigma}{\omega\epsilon_0} = (n - jk)^2 = 1 + \frac{Ne^2/m\epsilon_0}{-\omega^2 + j\omega g} + \sum_k{}' \frac{N_k e^2/m\epsilon_0}{\omega_k^2 - \omega^2 + j\omega g_k}, \quad (3.2)$$

in which it is clear that the term coming from the conductivity has just the same form as the other terms, except for having its resonant frequency equal to zero. Separating real and imaginary parts of (3.2), and using (3.1), where we shall now use σ for its value at zero frequency, we may rewrite (3.2) in the form

$$n^2 - k^2 = 1 - \frac{\sigma}{g\epsilon_0} \frac{1}{1 + \omega^2/g^2} + \sum_k{}' \frac{(N_k e^2/m\epsilon_0)(\omega_k^2 - \omega^2)}{(\omega_k^2 - \omega^2) + \omega^2 g_k^2}$$

$$2nk = \frac{\sigma}{\omega\epsilon_0} \frac{1}{1 + \omega^2/g^2} + \sum_k{}' \frac{(N_k e^2/m\epsilon_0)(\omega g_k)}{(\omega_k^2 - \omega^2)^2 + \omega^2 g_k^2}.$$

We notice that as the frequency becomes low compared with σ/ϵ_0, which for good conductors is in the ultraviolet part of the spectrum, the first term in the product nk becomes large compared with unity, masking the effect of the bound electrons. The difference $n^2 - k^2$ does not become correspondingly large, so that in the limit of low frequencies, n becomes equal to k, and both approach $\sqrt{\sigma/2\omega\epsilon_0}$, neglecting ω compared with g. However, it is only at low frequencies that these simplifications occur. As the frequency enters the near infrared or visible region, it becomes of the same order of magnitude as both σ/ϵ_0 and g, so that the contributions of the free electrons become complicated, and at the same time nk decreases so that the contributions of the bound electrons become important. It is thus natural that experimentally the curves of $n^2 - k^2$ and nk throughout the visible part of the spectrum are very complicated, though they can be fitted fairly accurately with formulas of the type we have derived, assuming bound as well as free electrons. In the ultraviolet, the contributions of the free electrons become small compared with those of the bound electrons having resonance in that region, and a metal does not behave essentially differently from an insulator.

4. The Quantum Theory and Dispersion.—The picture of dispersion and of the optical properties of dielectrics and metals that we have presented is based on simple classical models of the behavior of electrons in these materials. We have assumed that dielectrics contain electrons held to positions of equilibrium by linear restoring forces, subject to resistances proportional to their velocity, and we have assumed the conduction electrons in a metal to be similar, except that they have no restoring forces, but only resistance. The theory in this form was developed by Drude and Lorentz, in the early days of the electron theory. It has proved in practice to be so good as to surprise one, when the crudity of the assumptions is considered. The real electrons in atoms, molecules, and metals, as we now know from the quantum theory, behave very differently from our simple picture, but it is a remarkable fact that the quantum theory leads essentially to a justification of our mathematical formulation, though not of the simple hypotheses underlying it. According to the quantum theory, as is well known, atoms, molecules, and other systems have certain energy levels in which they can exist, and the emission and absorption of radiation are associated with transitions between energy levels, the frequency ν associated with two levels E_1 and E_2 being given by Planck's relation $E_2 - E_1 = h\nu$, where h is Planck's constant. Atoms and molecules have only a discrete set of excited energy levels, so that from the ground state they can absorb a discrete set of definite frequencies. Metals, on the other hand, have a continuum of excited levels, associated with the free electrons.

In the quantum theory, we can investigate the behavior of an atom, molecule, or metal, under the action of an external radiation field. We find, by application of certain perturbation methods, that the effect of the atom on the field can be replaced by an equivalent set of linear oscillators, associated with the various transitions that are allowed by Planck's relation. Each oscillator has a resonant frequency given by one of the allowed transitions. With N atoms per unit volume, however, we do not have the equivalent of N oscillating electrons per unit volume of each of the frequencies; we have instead only Nf, where f is a fraction, sometimes called the "oscillator strength," and which can be calculated by quantum mechanics. We further find that the oscillators have not only resonant frequencies, but also damping terms, which can be correlated with the broadening of the upper and lower states, by collisions with other atoms and other perturbing processes. Thus the net result of the quantum discussion is a theory mathematically equivalent to the one developed in the present chapter, but with quite a different physical interpretation. It is for this reason that a purely classical discussion of dispersion, such as we have given, is of real physical importance, and not merely an academic matter.

In metals, the quantum theory also leads to quite a different picture from an elementary classical theory. Electrons are not governed by classical statistics, but by Fermi statistics, as a result of which the electrons in a metal are not at rest in the absence of a field, as our classical equation of motion would suggest, but actually are in continuous motion with a very high velocity. In the absence of an external field, however, as many are traveling in one direction as in the opposite direction, so that there is no net current. In the presence of a field, the electrons gain an average acceleration that is essentially what corresponding classical electrons would experience, though in most cases they have an effective mass that is different from their classical mass, because of the structure of the energy bands of the metal. The electrons, once they are accelerated, are subject, in quantum theory as in our classical picture, to a frictional resistive force, but we can give a physical explanation of the friction, instead of merely postulating it. The electrons have many of the properties of waves, as is observed experimentally in electron diffraction, and these waves are scattered by the irregularities in the metal produced by the thermal oscillations of the atoms. It is this scattering which produces the effect of friction, dissipating any average momentum that the electrons may acquire. Thus the quantum theory eventually arrives at a picture of the electrons in a metal that is not unlike that of elementary classical theory, and the quantum theory of the interaction of metals with a radiation field essentially verifies the simple theory that we have described in this chapter.

This brief discussion of the relation of the classical theory of dispersion to the quantum theory is not expected to give a clear idea of the quantum phenomena to one not already familiar with them, but is intended merely to make it plain that, as we have already mentioned, the classical theory of dispersion is of great importance physically, furnishing a description of experimental facts that in its broad outlines is correct, and correlates with the quantum theory.

Problems

1. Show that, in the case of normal dispersion for the visible spectrum where there is an absorption band in the ultraviolet, the index of refraction can be written as

$$n^2 = A + \frac{B}{\lambda^2} + \frac{C}{\lambda^4} + \cdots,$$

where λ is the wave length in vacuum and A, B, C are constants.

If there is also absorption in the infrared, show that the index of refraction is then given by

$$n^2 = A + \frac{B}{\lambda^2} + \frac{C}{\lambda^4} + \cdots - A'\lambda^2 - B'\lambda^4 \cdots.$$

2. Measurements of H_2 gas give the following values of the index of refraction:

λ in A	$(n-1)10^7$
5,462.260	1,396
4,078.991	1,426
3,342.438	1,461
2,894.452	1,499
2,535.560	1,547
2,302.870	1,594
1,935.846	1,718
1,854.637	1,760

Using the expression in Prob. 1 for n^2 in reciprocal powers of λ, calculate the best values of A, B, and C. If the measurements are made at room temperature and atmospheric pressure, calculate the resonant frequency ω_0 and wave length from these constants.

3. Prove that in the case of anomalous dispersion for gases the maximum and minimum values of n occur at the positions where the absorption coefficient reaches half its maximum value. Find the relation between g_k and the half width of the absorption band. Assume $g_k/\omega_k \ll 1$.

4. For the D line of sodium the following values of the constants in the dispersion formula are found:

$$\omega_0 = 3 \times 10^{15}; \qquad g = 2 \times 10^{10}; \qquad \frac{Ne^2}{m\epsilon_0} = 10^{23}.$$

Plot the index of refraction n and the absorption coefficient k as a function of the frequency of the light. Find the maximum and minimum values of the index of refraction n. Find the maximum value of the absorption coefficient k and the half width of the absorption band in angstrom units.

5. Show that for gases the Lorenz-Lorentz law takes the approximate form $\frac{2}{3} \frac{n-1}{\rho_0} = $ constant. The following measurements have been made on air (ρ_0 given in arbitrary units)·

ρ_0	n
1.00	1.0002929
14.84	1.004338
42.13	1.01241
69.24	1.02044
96.16	1.02842
123.04	1.03633
149.53	1.04421
176.27	1.05213

Calculate $\frac{2}{3} \frac{n-1}{\rho_0}$ and $\frac{n^2-1}{n^2+2} \frac{1}{\rho_0}$ for each of these measurements, and compare the constancy of the results (calculate to four significant figures).

6. The indices of refraction for the sodium D line, and densities in grams per cubic centimeter of some liquids at 15°C are

Liquid	ρ_0	n
Water..................................	0.9991	1.3337
Carbon bisulphide.......................	1.2709	1.6320
Ethyl ether.............................	0.7200	1.3558

Calculate the indices of refraction for the vapors at 0°C and 760 mm pressure. The observed values for the vapors are 1.000250, 1.00148, and 1.00152, respectively.

7. The quantity $\dfrac{m(n^2 - 1)}{(n^2 + 2)\rho_0}$ is called the "refractivity" of a substance if m denotes its mass. Prove that the refractivities of mixtures of substances equal the sum of the refractivities of the constituents. (Neglect damping forces from the start.)

8. Show that the molecular refractivity of a compound, defined as $\dfrac{n^2 - 1}{n^2 + 2}\dfrac{M}{\rho_0}$, where M is the molecular weight, is equal to the sum of the atomic refractivities of the atoms of which the compound is formed. (Neglect damping forces.)

9. For the following gases we have the following values of $(n - 1)_\infty$ extrapolated to long wave lengths:

Gas	$(n - 1)_\infty \cdot 10^6$
H_2...	136.35
N_2...	294.5
O_2...	265.3

Calculate the values of $(n - 1)_\infty$ for the following gases: H_2O, NH_3, NO, N_2O_4, O_3. The measured values are 245.6, 364.6, 288.2, 496.5, 483.6, all times 10^6. Find the percentage discrepancy between the calculated and observed values.

CHAPTER X

REFLECTION AND REFRACTION
OF ELECTROMAGNETIC WAVES

In Chap. VIII we investigated the behavior of a plane wave in a homogeneous medium, and showed that it is propagated with a velocity equal to the velocity in free space, divided by the index of refraction. Next in Chap. IX we investigated the physical nature of various types of media, and found the features leading to various indices of refraction, their variation with wave length, and to the absorption that often accompanies dispersion. Now we shall consider what happens at an infinite plane boundary between two semi-infinite media of different indices of refraction, such for instance as free space and a dielectric, or free space and a metallic conductor. We shall be led to the familiar laws of reflection and refraction, laws that were established, before the electromagnetic theory, merely from general wave theory. To consider the behavior at the boundary, we must find the boundary conditions holding at a surface of discontinuity between two media.

1. Boundary Conditions at a Surface of Discontinuity.—In each of our two media, we assume that the solution of Maxwell's equations that we desire is a plane wave, just as in an infinite medium. At the boundary between the media, however, certain boundary conditions are to be met, and these demand that there be definite relations between the waves in the two media. These conditions have been derived earlier, in Eq. (3.3) of Chap. IV for the electric vectors, and in Eq. (3.2) of Chap. VI for the magnetic ones. We rewrite them:

> Normal components of D and B are continuous,
> Tangential components of E and H are continuous,   (1.1)

at a surface that contains no charge and current. These conditions will prove to be enough to derive the complete laws of reflection and refraction; in fact, the conditions on D and B follow from the others, and are not necessary for the derivation. We shall find that in general we cannot satisfy them without postulating three separate plane waves: an incident and a reflected wave in one medium, a refracted wave in the other medium. By arguments like those of elementary optics, involving just the matching up of wave fronts on both sides of the surface of separation, we can prove the simple laws of reflection and refraction: that the angles of incidence and reflection are equal, and Snell's law of refraction. We can go further than this, however, and compute the intensities of the reflected and refracted waves, and hence the reflection coefficients, embodied in Fresnel's equations.

We shall first prove the laws of reflection and refraction, and then Fresnel's equations. Then we shall take up two more complicated cases, which cannot be handled by the most elementary methods: total internal reflection, in which there is an exponentially damped wave of an interesting type in the rarer medium, which does not lead to a flow of power into the medium; and reflection and refraction by a conducting medium such as a metal, in which Snell's law of refraction does not hold in its simple form. In all this discussion, we notice that our treatment holds for any part of the spectrum with equal validity, from the longest electromagnetic or radio waves, through the visible region, to the X rays and gamma rays. The only difference is that the index of refraction and absorption coefficient vary with wave length, in the manner we took up in the preceding chapter. These quantities will appear as constants in our present discussion, however, which will deal throughout with monochromatic waves, of a definite wave length.

2. The Laws of Reflection and Refraction.—Let us assume that the surface of separation is $z = 0$, and that the medium with negative z has constants μ, ϵ, and that for positive values has μ', ϵ', with corresponding indices of refraction and absorption coefficients n, k, and n', k'. Let the incident wave be in the medium with negative z, and let its wave normal have direction cosines l_1, l_2, l_3. Then, in a non-absorbing medium, which alone we consider in the present section, the exponential factor representing the wave propagation is

$$e^{j\omega[t-(l_1 x+l_2 y+l_3 z)/v)]}, \tag{2.1}$$

where $v = c/n$ is the velocity of propagation. We shall simplify by assuming that the wave normal is in the xz plane, and that the angle of incidence, or angle between the wave normal and the z axis, is i; then we have $l_1 = \sin i$, $l_2 = 0$, $l_3 = \cos i$, so that (2.1) may be rewritten

$$e^{j\omega[t-(x \sin i+z \cos i)/v]}. \tag{2.2}$$

Similarly in the medium with positive z, there will be a wave propagated at an angle r with the normal, r being the angle of refraction; its corresponding exponential factor is

$$e^{j\omega[t-(x \sin r+z \cos r)/v']}. \tag{2.3}$$

Along the surface of separation, $z = 0$, the two exponentials must agree; for otherwise, even if we were able to satisfy boundary conditions at one point, these would not hold at other points of the surface. Thus we must have

$$\frac{\sin i}{\sin r} = \frac{v}{v'} = \frac{n'}{n},\qquad(2.4)$$

which is Snell's law of refraction. Our derivation is simply the analytical statement of the elementary fact that the wave fronts in the two media must match each other. In addition to the refracted wave, we can set up in the first medium a reflected wave, whose z component of propagation is reversed, with the exponential

$$e^{j\omega[t-(x\,\sin\,i-z\,\cos\,i)/v]}.\qquad(2.5)$$

Along the plane $z = 0$, this exponential agrees with the other two, so that the boundary conditions can be satisfied; and a little consideration shows that it satisfies the law of reflection, that the angle of reflection equals the angle of incidence i.

3. Reflection Coefficient at Normal Incidence.—We shall next set up the values of the electric and magnetic vectors on both sides of the boundary, use the boundary conditions (1.1), and derive the relations between the amplitudes of incident, reflected, and refracted waves. As a preliminary simple case, let us consider the case of normal incidence, when i and r are zero, so that the exponentials (2.2), (2.3), and (2.5) are independent of x. There will be no normal components of **D** and **B,** since the fields are transverse to the direction of propagation, which in this case is the z axis. In the incident wave, let **E** be along the x axis, **H** along the y axis. From Eq. (2.1) of Chap. VIII, we have

$$\frac{E_x}{H_y} = Z_0 = \sqrt{\frac{\mu}{\epsilon}}.\qquad(3.1)$$

Similarly for the refracted wave we assume **E** is along x, **H** along y, and have

$$\frac{E'_x}{H'_y} = Z'_0 = \sqrt{\frac{\mu'}{\epsilon'}}.\qquad(3.2)$$

For the reflected wave, either **E** or **H** must be reversed in phase with respect to the incident wave, since the wave travels in the opposite direction. Thus we have

$$\frac{E''_x}{H''_y} = -\sqrt{\frac{\mu}{\epsilon}}.\qquad(3.3)$$

Our boundary conditions may then be stated with respect to $\mathbf{E}$:

$$E_x + E_x'' = E_x', \qquad \sqrt{\frac{\epsilon}{\mu}}\,(E_x - E_x'') = \sqrt{\frac{\epsilon'}{\mu'}}\,E_x'.$$

Eliminating E_x', we find for the ratio of reflected to incident amplitude

$$\frac{E_x''}{E_x} = \frac{\sqrt{\epsilon/\mu} - \sqrt{\epsilon'/\mu'}}{\sqrt{\epsilon/\mu} + \sqrt{\epsilon'/\mu'}} = \frac{Z_0' - Z_0}{Z_0' + Z_0}. \tag{3.4}$$

The ratio of reflected H to incident H, except for the change of sign, is the same; thus the ratio of reflected power to incident power, or the reflection coefficient for power, is the square of the quantity given in (3.4).

In most common cases, the media are nonmagnetic, and

$$\mu = \mu' = \mu_0,$$

the value characteristic of free space. In this case, from (1.9), Chap. VIII, the square roots in (3.4) are proportional to the corresponding indices of refraction, and the reflection coefficient for power is

$$\text{Reflection coefficient for power} = \frac{(n - n')^2}{(n + n')^2}. \tag{3.5}$$

We note that, if n and n' are interchanged, or the incident wave is approaching the surface from the other side, the reflection coefficient is unchanged; a surface always reflects equally in both directions. The reflection coefficient is of course less than unity. We can easily find E_x', the amplitude of the refracted wave, if we choose, and show that the power carried to the surface by the incident wave is split up between the reflected and refracted waves, so that, if any energy is in the refracted wave, the reflected wave must be weaker than the incident, and the reflection coefficient must be less than unity. For instance, for a wave of light reflected at a surface between glass and air, we have $n = 1$ for air, and about 1.5 for glass; in this case the coefficient is $(0.5)^2/(2.5)^2 = \frac{1}{25}$, showing that only 4 per cent of the intensity is reflected from a glass plate at normal incidence.

4. Fresnel's Equations.—Next we take the case of an arbitrary angle of incidence. Here we meet the question of polarization. The vector $\mathbf{E}$ is at right angles to the direction of propagation, but that does not fix the direction uniquely, and it is said that the wave is polarized in a particular direction if its electric vector (or, according to an alternative convention, its magnetic vector) points in that direction. Let us then consider the two extreme cases. We take the wave normal of the incident wave to be in the xz plane, as before. Then we consider the case where the electric vector is along the y axis, and the case where it is in the xz plane, as in Fig. 25. In case 1, the y axis points down into the paper, $\mathbf{E}$ and $\mathbf{E}'$ point down, and $\mathbf{E}''$ points up; in case 2, $\mathbf{H}, \mathbf{H}'$, and $\mathbf{H}''$ all point down, along the y axis. We now discuss these cases separately.

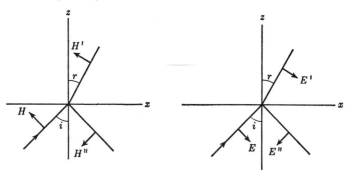

FIG. 25.—Vectors in reflection and refraction. Case 1: y axis points down into the paper. E and E' point down, E'' points up. Case 2: H, H', H'' all point down.

Case 1.—The electric vector is along the y axis, or at right angles to the plane of incidence. All vectors depend on space in the way indicated by (2.2) for the incident wave, (2.3) for the refracted wave, and (2.5) for the reflected wave. From the figure, we see that for the incident wave $H_x = -(E_y/Z_0) \cos i$, where Z_0 is given in (3.1). Similarly for the refracted wave, $H_x' = -(E_y'/Z_0') \cos r$, and for the reflected wave $H_x'' = -(E_y''/Z_0) \cos i$. Hence we have the following relations:

Tangential component of $\mathbf{E}$: $E_y - E_y'' = E_y'$.

Tangential component of $\mathbf{H}$: $-(E_y + E_y'') \dfrac{\cos i}{Z_0} = -E_y' \dfrac{\cos r}{Z_0'}$.

Remembering Snell's law, the second may be rewritten

$$E_y + E_y'' = E_y' \frac{\mu}{\mu'} \frac{\tan i}{\tan r}.$$

From this at once, multiplying the first by $(\mu/\mu')(\tan i/\tan r)$, and subtracting, we have

$$E_y \left(\frac{\mu}{\mu'} \frac{\tan i}{\tan r} - 1 \right) = E_y'' \left(\frac{\mu}{\mu'} \frac{\tan i}{\tan r} + 1 \right),$$

$$\frac{E_y''}{E_y} = \frac{\mu \tan i - \mu' \tan r}{\mu \tan i + \mu' \tan r}.$$

If both media are nonmagnetic, so that $\mu = \mu'$, as is usually the case, this becomes

$$\frac{E_y''}{E_y} = \frac{\tan i - \tan r}{\tan i + \tan r} = \frac{\sin i \cos r - \cos i \sin r}{\sin i \cos r + \cos i \sin r},$$

$$\frac{E_y''}{E_y} = \frac{\sin (i - r)}{\sin (i + r)}. \tag{4.1}$$

This gives the amplitude of the reflected wave, and is one of Fresnel's equations. We note that, as i and r become zero, the law of refraction becomes $i/r = n'/n$, $i = (n'/n)r$. Thus, in the limit of normal incidence, the ratio approaches $(n' - n)/(n' + n)$, as we found above (when we correct the signs for the different convention used here). We also note, in the other extreme of tangential or grazing incidence, that $i = 90°$, so that the ratio is $[\sin (90° - r)]/[\sin (90° + r)] = 1$. That is, the reflection coefficient equals unity for grazing incidence. The formula gives a monotonic increase of amplitude as the angle of incidence increases.

Case 2.—The electric vector is in the xz plane, or in the plane of incidence. We let **H** be along the y axis in all waves, and let E, E', E'' be the magnitudes of **E** in the three waves. Then $H_y = E/Z_0$, $H_y' = E'/Z_0'$, $H_y'' = E''/Z_0$. For the components of the electric vectors, we have $E_x = E \cos i$, $E_x' = E' \cos r$, $E_x'' = -E'' \cos i$. Then we have

Tangential component of **E**: $(E - E'') \cos i = E' \cos r$

Tangential component of **H**: $\dfrac{(E + E'')}{Z_0} = \dfrac{E'}{Z_0'}$.

We may rewrite the first as $E - E'' = E' (\cos r/\cos i)$, the second as $E + E'' = (Z_0/Z_0')E'$. Multiplying the first by

$$\frac{Z_0}{Z_0'} = \left(\frac{\mu}{\mu'}\right)\left(\frac{n'}{n}\right) = \left(\frac{\mu}{\mu'}\right)\left(\frac{\sin i}{\sin r}\right),$$

the second by $\cos r/\cos i$, and subtracting, we have

$$E\left(\frac{\mu}{\mu'}\frac{\sin i}{\sin r} - \frac{\cos r}{\cos i}\right) = E''\left(\frac{\mu}{\mu'}\frac{\sin i}{\sin r} + \frac{\cos r}{\cos i}\right),$$

$$\frac{E''}{E} = \frac{\mu \sin i \cos i - \mu' \sin r \cos r}{\mu \sin i \cos i + \mu' \sin r \cos r}.$$

If $\mu = \mu'$, we can use the trigonometric relations

$\sin (i \pm r) \cos (i \mp r)$

$$= (\sin i \cos r \pm \cos i \sin r)(\cos i \cos r \pm \sin i \sin r)$$
$$= \sin i \cos i(\cos^2 r + \sin^2 r) \pm \sin r \cos r(\sin^2 i + \cos^2 i)$$
$$= \sin i \cos i \pm \sin r \cos r.$$

Hence we have

$$\frac{E''}{E} = \frac{\sin (i - r) \cos (i + r)}{\sin (i + r) \cos (i - r)} = \frac{\tan (i - r)}{\tan (i + r)}. \tag{4.2}$$

This is the other of Fresnel's equations.

There is one interesting feature met in this case, which is not present when the electric vector is at right angles to the plane of incidence: if $i + r = 90°$, a perfectly possible situation, the denominator of (4.2) is infinite, so that the reflection coefficient is zero. This angle is called the "polarizing angle"; if unpolarized radiation, consisting of a mixture of both types of radiation, falls on a surface at this angle, only the radiation with its electric vector at right angles to the plane of incidence will be reflected, and the reflected radiation will be polarized. It was by this phenomenon that polarized light was first discovered. Light was reflected from one mirror at this angle. Then its polarization was found by reflecting from a second mirror at the same angle. As the second mirror was rotated about the beam as an axis, so that the polarization changed from being at right angles to the plane of incidence to being in the plane, the doubly reflected beam changed from a maximum intensity to zero. The polarizing angle r' is fixed by $i' + r' = 90°$, and this occurs when $\cos i' = \sin r'$. Using the law of refraction, we find

$$\tan i' = \frac{n'}{n},$$

thus fixing the polarizing angle i'.

5. Total Reflection.—For radiation passing from a dense medium to a rarer medium, so that $n' < n$, the angle of refraction becomes $90°$ for an angle i for which $\sin i = n'/n$. For greater angles of incidence, the equation $\sin r = (n/n') \sin i$ would indicate that $\sin r$ should be greater than unity, so that r must be imaginary. To find the physical meaning of this situation, we compute $\cos r$, which is given by

$$\cos r = \pm \sqrt{1 - \sin^2 r} = \pm j \sqrt{(n/n')^2 \sin^2 i - 1}.$$

The expression (2.3) for the disturbance in the rarer medium then becomes

$$e^{i\omega[t - (x \sin r/v')]}e^{-\omega[\sqrt{(n/n')^2 \sin^2 i - 1}]z/v'},$$

where we have used the negative square root. The first term represents a wave propagated along the x axis, or parallel to the surface of the medium, with an apparent velocity $v'/\sin r$, a value less than v'. The second factor indicates that the amplitude of this wave is damped out as z increases, or as we go away from the surface, so that the wave fronts (surfaces of constant phase) are at right angles to the surfaces of constant amplitude. This disturbance ordinarily damps out in a very short distance. Thus if $(n/n') \sin^2 i$ is decidedly greater than 1, the exponential becomes small when z is a few wave lengths ($\omega z/v'$ a reasonably large number). Consequently the disturbance in the rare medium is not observed, unless special experiments are devised to find it. It is easily shown that Poynting's vector for this wave has no component normal to the surface, so that it does not carry any energy away. Thus all the energy in the incident wave must reappear in the reflected wave, and for this reason the phenomenon is called "total reflection." The angle of incidence given by $\sin i = n'/n$, which must be exceeded in order to have total reflection, is called the "critical angle."

Although there is no change of amplitude on reflection in this case, there is a change of phase, which is sometimes of interest, and which may be treated by Fresnel's equations. Thus in case 1 we have, assuming nonmagnetic media,

$$\frac{E''}{E} = \frac{\sin i \cos r - \cos i \sin r}{\sin i \cos r + \cos i \sin r} = \frac{\sqrt{\sin^2 i - (n'/n)^2} - j \cos i}{\sqrt{\sin^2 i - (n'/n)^2} + j \cos i}$$

and in case 2,

$$\frac{E''}{E} = -\frac{(n/n')^2 \sqrt{\sin^2 i - (n'/n)^2} - j \cos i}{(n/n')^2 \sqrt{\sin^2 i - (n'/n)^2} + j \cos i}$$

In each case, E''/E is the ratio of a complex number to its complex conjugate, which is therefore a complex number of unit magnitude (showing that the reflection coefficient is unity), but with a certain phase angle. Thus, in the general case, where E has components both in the xz plane and along the y axis, there is a difference of phase between these components upon total reflection, and linearly polarized light in general will become elliptically polarized upon total reflection. To see this, we note that two vibrations at right angles, with the same frequency and phase, produce a resultant vector whose extremity moves in a line (plane polarization), but if the two components are in different phases the extremity of the vector traces out an ellipse. If the phases differ by 90°, and the amplitudes are equal, the polarization is circular.

6. Damped Plane Waves, Normal Incidence.—So far, we have taken up only the case of nonabsorbing media, with real indices of refraction. The case of reflection by an absorbing medium, such as a metal, can be handled, as we saw in Chap. IX, by replacing the index of refraction n by the complex quantity $n - jk$. Considering the case of normal incidence, using (3.4), and setting $\mu = \mu'$, we have

$$\frac{E''_x}{E_x} = \frac{n - (n' - jk)}{n + (n' - jk')} = \frac{(n - n') + jk'}{(n + n') - jk'}, \tag{6.1}$$

on the assumption that the wave is incident on an absorbing medium (n',k') from a nonabsorbing medium. Since this ratio is complex, we see in the first place that there is a phase change on reflection from an absorbing medium. To find the reflection coefficient R, we must take the square of the magnitude of the quantity in (6.1). This can be done by multiplying by the complex conjugate. We have

$$R = \frac{(n - n')^2 + k'^2}{(n + n')^2 + k'^2}.$$

It is interesting to consider the behavior of a metal at relatively low frequencies. We saw in Chap. IX, Sec. 3, that at frequencies below the shorter part of the infrared, n' and k' both approach $\sqrt{\sigma/2\omega\epsilon_0}$. Furthermore, both these quantities are very large compared with unity, or with n, the index of refraction of the transparent medium in question. Substituting these values, the reflection coefficient becomes

$$R = 1 - 2n\sqrt{\frac{2\omega\epsilon_0}{\sigma}}. \tag{6.2}$$

According to our assumptions, the second term is small; thus nearly all the incident radiation is reflected. Furthermore, the second term becomes smaller as the frequency is reduced, or as the conductivity increases. In these cases, very little power flows into the conductor, and is dissipated in it because of the damping of the wave, or, in physical language, because of the dissipation of heat in the resistance. The reason so little power flows into the metal is easily seen, from the relation (2.1) of Chap. VIII for the ratio of E to H. From that equation, in the limit of low frequency, we have

$$\frac{E_x}{H_y} = \sqrt{\frac{j\mu\omega}{\sigma}} = \sqrt{\frac{\mu_0}{\epsilon_0}}\sqrt{\frac{j\omega\epsilon_0}{\sigma}},$$

if we assume that $\mu = \mu_0$. That is, the ratio of E to H is much smaller than the value $\sqrt{\mu_0/\epsilon_0}$ characteristic of empty space. It is so small, in fact, that the E within the metal can almost be neglected. Thus

Poynting's vector at the surface is very small, representing a small energy flow into the metal. The situation is almost like that with a perfect conductor, in which the electric vectors of the incident and reflected waves exactly cancel at the surface of the metal, whereas the magnetic vectors are equal in magnitude, because of the perfect reflection, and add.

Looking back to the value of γ for a metal, from Eq. (1.5) of Chap. VIII, we see that in our limiting case

$$\gamma = \pm j\omega \sqrt{-\frac{j\sigma\mu}{\omega}} = \pm j \sqrt{-j\omega\sigma\mu}.$$

This tells us in the first place that the real and imaginary parts are equal [since $j \sqrt{-j} = (1 + j)/\sqrt{2}$], so that a wave inside a good conductor damps down to a small fraction of its intensity in a few wave lengths; and that the distance in which the intensity falls off to a fraction of its value at the surface decreases as either the frequency or conductivity increases. It is often convenient to define the distance in which the amplitude falls to $1/e$ of its value as the skin depth, for this damping of the wave inside the conductor is simply another way of describing the skin effect, familiar in high-frequency electric-circuit theory. We have for δ, the skin depth,

$$\delta = \sqrt{\frac{2}{\omega\sigma\mu}}. \tag{6.3}$$

For the limiting case of a perfect conductor, we then see that the reflection coefficient is unity; the tangential E at the surface is zero (as it must be, for otherwise the current would have to be infinite), but the tangential H is finite; but that the field within the conductor is a damped wave that falls off infinitely rapidly to zero as we penetrate the metal, so that directly below the surface H, as well as E, is zero. In other words, to take account of the rapid decrease of the tangential H from a finite value at the surface to zero directly below the surface, Stokes's theorem shows that we must assume a surface-current density at the surface of the conductor, numerically equal to the tangential H, but at right angles to it, which because of the perfect conductivity can flow without a corresponding tangential electric field. These limiting boundary conditions are often convenient to use directly in discussing the boundary conditions of an electromagnetic wave reflected by a perfect conductor, as for instance in considering propagation in wave guides, and electromagnetic fields in resonant microwave cavities.

7. Damped Plane Waves, Oblique Incidence.—If a plane wave approaches the surface between a transparent medium and an absorbing medium, such as a metal, at oblique incidence, the problem is considerably more complicated than any that we have taken up so far, and we shall not give a complete discussion. (A more complete treatment is given in J. C. Slater, *Microwave Transmission*, Sec. 13, Dover Publications, Inc., New York, 1959. We can see the reason for the complication in a very simple way: The wave normal in the metal will be at an angle of refraction r, which certainly will not bring it along the z axis, or the normal to the surface. Thus the surfaces of constant phase, or the wave fronts, in the metal, will not be parallel to the planes z = constant. On the other hand, the amplitude is constant over the surface $z = 0$, the surface of the metal; thus it will also be constant at surfaces z = constant within the metal, since at all points of such a surface the wave will have penetrated equal distances through the absorbing medium, and will have had equal energy losses. Thus the surfaces of constant phase do not coincide with the surfaces of constant amplitude.

We have already seen, in Sec. 5, an example of a case where these surfaces do not coincide; in that case, in fact, they were at right angles. We may get a general approach to this problem by assuming, in place of (1.4) of Chap. VIII, a general solution of the wave equation of the form $e^{j\omega t - \gamma \cdot r}$, where γ and r are now vectors, and where γ can be called the "propagation vector." In Chap. VIII, we could have used this form, with the assumption that γ pointed along the z axis. We saw in (1.6) of Chap. VIII that γ could be complex; but now we see that an additional complication can arise, in that the real and imaginary parts of γ, which we can call α and β, can be vectors that do not have to have the same direction in space. If we substitute our expression in the wave equation, we find that, if there is no loss in the medium, α and β must be at right angles to each other, and there is a definite relation between their magnitudes. This is the situation met in total reflection, where we have a damped wave, but in a nonabsorbing medium. The physical situation is clear: since there is no loss, the intensity of the wave must be constant along the direction of propagation, but there is nothing to prevent its intensity varying at right angles to this direction.

On the other hand, in an absorbing medium, the relation between the directions of α and β is quite arbitrary. Thus we can always choose the direction of α so as to make the surface of the metal a surface of constant amplitude, and yet can choose the direction of β

so as to satisfy the law of refraction. We find, however, that the law of refraction is no longer the simple form of Snell's law. The angle of refraction acts like a complex angle, and we can modify Snell's law, if we choose, to give the correct results, by taking a complex index of refraction, and complex angle of refraction. The net result of the law of refraction, however, is very simple for a metal of good conductivity. We have seen that its index of refraction n', as well as the absorption coefficient k', are very large. The refracted wave within the metal resembles that in an ordinary case of refraction to the extent that, with a very large index of refraction, the angle of refraction is very small, or the wave normal of the refracted wave is nearly along the normal to the surface. In other words, quite independent of the angle of incidence, the wave inside the metal is similar to that discussed in Sec. 6. There is, however, a change of phase on reflection, which we can get from Fresnel's equations, treating the angle of refraction as complex, which is a function of the angle of incidence, and which is different for the two planes of polarization of the incident wave. Thus it can happen that the state of polarization of the incident wave can be changed by reflection, in a characteristic manner, and this gives one of the practical experimental methods for finding information regarding the optical constants of a metal.

Problems

1. Light is reflected from glass of index of refraction 1.5. Compute and plot curves for the reflected intensity as a function of angle, for both sorts of plane polarization.

2. Find the intensity of light in the refracted medium, for arbitrary angle of incidence and both types of polarization. Show that the amount of energy striking the surface is just equal to the amount carried away from it. Note that the amount striking the surface is computed, not from the whole of Poynting's vector, but from its normal component.

3. Light passes normally through a glass plate. Find the weakening in intensity because of the reflection at the faces.

4. Ten plates of glass of index 1.5 are placed together and used as a polarizer. Light strikes the plates at the polarizing angle, and the transmitted light is used. Since all the reflected light is of one polarization, and the reflection at both surfaces of all plates is enough to remove practically all the light of this polarization, the transmitted light will be practically polarized in the other direction. Find the intensity of both sorts of light in the transmitted beam, assuming initially unpolarized light, and hence show how much polarization is introduced. You may have to consider multiple internal reflection.

5. The resistivity of copper is about 1.7×10^{-6} ohm-cm. Calculate the reflective power of copper for wave lengths of light $\lambda = 12\mu$ and $\lambda = 25.5\mu$. The observed values of $1 - R$ are 1.6 per cent and 1.17 per cent at these wave lengths.

CHAPTER XI

WAVE GUIDES AND CAVITY RESONATORS

If electromagnetic radiation is introduced into the open end of a hollow pipe bounded by perfectly conducting, and hence perfectly reflecting walls, the radiation will be reflected from the wall whenever it strikes it, and will be able to progress down the pipe for an indefinite distance. Such a hollow pipe is called a "wave guide," and has recently come into much prominence in connection with the applications of microwaves, or electromagnetic waves whose lengths are of the order of magnitude of laboratory dimensions. The propagation of waves down a wave guide is not quite so simple as it seems at first sight; for the various reflected waves interfere with each other, in such a way that only certain types of waves can exist in the guide. In particular, the field pattern of the wave in a plane perpendicular to the axis can take on only one of a discrete, though infinite, number of character-istic patterns, called "modes," determined by the shape and size of the cross section. Corresponding to each of these modes, a wave is possible in the guide for any given frequency, and the effective wave length as measured along the guide, the so-called "guide wave length," is determined in terms of the frequency.

For each frequency and guide wave length we can at once deter-mine the phase velocity of propagation along the axis of the guide, and we find that in most cases this phase velocity is greater than the velocity of light, a fact that does not contradict the principle of rela-tivity, since it can be shown that no signal can be propagated with this phase velocity. The phase velocity varies with frequency; that is, the propagation shows dispersion, just like the propagation of light in a dispersive medium, which we discussed in Chap. IX. In fact, as the frequency decreases, the guide wave length, and the phase velocity, increase without limit, both becoming infinite at a quite finite fre-quency. At still lower frequencies, the disturbance, instead of being propagated along the wave guide, is exponentially attenuated or damped, at a more and more rapid rate as the frequency approaches zero. The frequency at which the disturbance changes over from being propagated to being attenuated is called the "cutoff frequency," and the corresponding wave length of a wave in free space is called the "cutoff wave length." Since the guide will propagate only waves of free-space wave length shorter than the cutoff wave length, or of frequencies higher than the cutoff frequency, it forms a filter, passing only high frequencies. The cutoff frequency is different for each mode. Thus, for frequencies less than the lowest cutoff frequency, no wave can be propagated down the guide; between the lowest and next lowest cutoff frequencies, only one mode can be propagated; and so on. As a practical matter, wave guides are usually used in the region between the lowest and next lowest cutoff frequencies, so as to have only one mode propagated, the rest attenuated. For certain special types of wave guides, the lowest cutoff frequency is zero, so that in such cases any frequency, no matter how low, can be propagated. We shall find that in such a case this mode of zero cutoff frequency, called the "dominant mode," has the interesting property that the velocity of propagation is equal to the velocity of light in free space, so that this mode, unlike the others, does not show dispersion.

If a length of wave guide is closed by two perfectly reflecting planes, radiation propagated down the guide will be reflected back and forth by the planes, and will form standing waves. Furthermore, in order that the successive reflections may reinforce each other, it is clearly necessary that the round trip, from one plane to the other and back again, should occupy a whole number of periods of the vibration, or that this distance should be a whole number of guide wave lengths. Thus the guide wave lengths, and hence the frequencies, are determined in terms of the dimensions. Such a guide with closed ends forms a simple example of a cavity resonator. It is found, in fact, that any closed region bounded by perfectly reflecting walls can vibrate in certain normal modes, each of a definite frequency fixed by the geometry of the cavity, there being an infinite number of such normal modes of greater and greater frequency. Such resonant cavities form the equivalent, in the microwave range of frequencies, of resonant circuits in the ordinary range where electric circuits are used. It is obvious that wave guides with closed ends, forming resonant cavities, form a close analogy to organ pipes, and it is not surprising to find that Lord Rayleigh, some of whose best known work was in the theory of sound, worked out the theory of wave guides and of cavity resonators in the latter part of the nineteenth century. Their practical application, however, has waited until the last few years, when techniques for producing the oscillations of the required very high frequency have become available.

The properties of wave guides, which we have outlined above, seem at first sight strange and unexpected, though it is very easy to prove these properties by setting up appropriate solutions of Maxwell's equations. A little further consideration, however, shows that the properties of wave guides follow largely from the discussion of reflection which we have already given. Even the attenuated waves for frequencies below cutoff prove to have their simple explanation, and to be closely related to the attenuated wave in the rarer medium in the problem of total internal reflection, the cutoff phenomenon having an analogy to the critical angle. Accordingly, so as to make the subject physically clear, we shall start our discussion by considering a very simple problem, the propagation of radiation between two parallel plane mirrors, which shows many of the properties met with wave guides. We shall not push this method too far, however, for in fact it is only with rectangular wave guides that one can build up the solution from a study of reflection by plane walls. In more complicated cases, such as the circular guide, this method is not available, and yet we can set up a general treatment that takes care of any arbitrary cross section in a fairly simple way.

1. Propagation between Two Parallel Mirrors.—A perfect reflector allows no field to penetrate within it; we have seen how this is to be realized in fact, by considering a good conductor, in which the field penetrates only to a distance comparable with the skin depth, which reduces to zero in the limit of perfect conductivity, which we assume. Thus the electric and magnetic fields are zero within the mirror, and our boundary conditions reduce to two statements: first, that the tangential component of **E** at the surface of the mirror is zero, since it must be zero within the mirror, and we know that the tangential component is continuous; secondly, the normal component of **B** at the surface must be zero, since **B** is zero within the mirror, and the normal component of **B** is continuous. The other two conditions, on the normal component of **D** and the tangential component of **H**, lead to information about the surface charge and current; for our surface is conducting and can carry charge and current, so that the value of the normal component of **D** outside the metal measures the surface-charge density, and the tangential component of **H** the surface-current density, which appear on the surface. Because of the lack of penetration of the field into the mirror, we need consider only incident waves striking the mirror, and corresponding reflected waves, but no refracted waves within the mirror.

With this understanding of the nature of the reflection process, let us consider a wave striking a perfectly reflecting mirror, and the corresponding reflected wave. Let the mirror be in the xz plane, the plane $y = 0$, and let the incident wave have its normal in the yz plane, making an angle of incidence i with the y axis, as shown in Fig. 26.

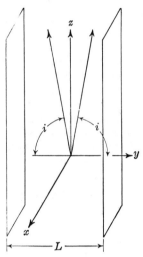

Let the incident wave approach from positive values of y, and negative z's. Then, using the same sort of arguments as in Chap. X, the exponential factor representing wave propagation in the incident wave will be

$$e^{j\omega\{t-[(-y\cos i+z\sin i)/c]\}} \qquad (1.1)$$

and the corresponding factor in the reflected wave will be

$$e^{j\omega\{t-[(y\cos i+z\sin i)/c]\}}. \qquad (1.2)$$

We must now superpose these in such a way that certain components of the field (the tangential component of **E**, and the normal component of **B**) will be zero on the plane $y = 0$. The only way in which we can accomplish this is to superpose (1.1) and (1.2) in such a way that the function of y turns into a sine function. We do this by subtracting (1.2) from (1.1), and dividing by $2j$. Thus we obtain the function

FIG. 26.—Coordinate system for propagation between parallel planes.

$$e^{j\omega[t-(z\sin i)/c]} \sin\left(\frac{\omega\cos i}{c}\right) y. \qquad (1.3)$$

The solution (1.3) may be interpreted as a disturbance that is propagated along the z axis with an apparent velocity $c/\sin i$, which is greater than c, since $\sin i$ is less than unity; and varying sinusoidally along y, or corresponding to standing waves along y, with an effective wave length given by

$$\frac{2\pi}{\lambda_c} = \frac{\omega\cos i}{c}, \qquad \lambda_c = \frac{\lambda_0}{\cos i}, \qquad (1.4)$$

where λ_c is the effective wave length along y, $\lambda_0 = 2\pi c/\omega$ is the free-space wave length, or wave length of the disturbance in the absence of reflections. The standing-wave phenomenon along y can be observed with visible light, in the form of the so-called "Lippmann fringes." Lippmann showed that, if standing waves are set up by this process, not in free space, but in the sensitive emulsion of a photographic plate, the emulsion will be blackened along surfaces half a wave length apart in the standing-wave pattern. The phenomenon can also be demonstrated very easily, with the longer waves of the microwave region.

We have so far considered only the reflection by one mirror, at $y = 0$. In case, however, we have a second mirror, at $y = L$, we must satisfy the same sort of boundary conditions at this second mirror. Thus our function (1.3) must be zero, not only at $y = 0$, but also at $y = L$. This leads at once to the condition

$$L = \frac{n\lambda_c}{2}, \tag{1.5}$$

where n is an integer. In other words, the dependence of the disturbance on y cannot have any arbitrary wave length, but only an infinite set of discrete wave lengths. These form the modes that we have already mentioned. It should be noted that, to discuss this problem of two mirrors, we do not have to think of multiple reflection, in the elementary sense. Our one wave traveling along $+y$, and the other traveling along $-y$, suffice to satisfy the boundary conditions everywhere. If we treat the disturbance by means of rays, we should say that any given ray is reflected back and forth an infinite number of times from the two mirrors, but in fact all the rays traveling to the right coalesce to furnish the description of the single wave traveling to the right, and similarly with the rays traveling to the left.

For a given n, or a given mode, the wave length λ_c is determined by (1.5). This is independent of the free-space wave length λ_0, or the corresponding frequency ω. Knowing λ_0, from the conditions of our problem, we can next use (1.4) to find i, the angle of incidence, and from this we can find the apparent phase velocity v_g along the z direction, determined by

$$v_g = \frac{c}{\sin i}, \tag{1.6}$$

and the corresponding wave length,

$$\lambda_g = \frac{v_g}{c}\lambda_0 = \frac{\lambda_0}{\sin i}. \tag{1.7}$$

For many purposes it is convenient to eliminate i, the angle of incidence. Computing $\cos i$ from (1.4), $\sin i$ from (1.7), and taking the sum of their squares, we find at once the relation

$$\frac{1}{\lambda_g^2} + \frac{1}{\lambda_c^2} = \frac{1}{\lambda_0^2}. \tag{1.8}$$

This simple equation determines the wave length λ_g along the z direction, in terms of the free-space wave length λ_0, and the wave length λ_c along the y direction, or the direction normal to z. We notice one interesting fact from (1.8). If λ_c is fixed, and if λ_0 is increased, or the frequency is decreased, $1/\lambda_g^2$ decreases, or λ_g increases, with consequent increase of the phase velocity v_g, until finally when λ_0 equals λ_c, λ_g and v_g become infinite. For values of λ_0 greater than λ_c, the wave length λ_g and the velocity v_g become imaginary, so that the wave is attenuated rather than propagated along the z axis. In other words, we have the phenomena that we have already described, including a cutoff wave length, with only attenuation for longer waves than the cutoff wave length. Furthermore, we see that λ_c is the cutoff wave length in question.

We have so far considered only the exponential function associated with the direct and reflected waves, and the corresponding relations regarding the wave length and frequency. In addition, we should take up the orientation and magnitude of the electric and magnetic vectors. As in Chap. X, there are two cases: that in which **E** is along the x direction, or at right angles to z for both the direct and reflected waves, and that in which **H** is in the x direction. In the language of wave guides, these are known, respectively, as the transverse electric (abbreviated TE) and transverse magnetic (abbreviated TM) modes of propagation. All wave guides, we shall find, have modes of both types. We could go through the equivalent of Fresnel's equations, and find the relations between **E** and **H** in these two types of modes, but it is so easy to prove much more general results holding for any type of guide, as we shall do in the next section, that we shall not carry out this discussion here.

Our problem of propagation between two mirrors forms a very simple and yet instructive case of wave-guide propagation. If we add two more reflecting surfaces perpendicular to the x axis, we have a rectangular wave guide. Some of the modes encountered with the mirrors, namely, those in which **E** is along the x axis, satisfy the boundary conditions for the rectangular guide as well, and in fact form important modes of this guide. They are not the whole set of modes, however, for there can be a sinusoidal variation along x as well as along y. If we furthermore add two reflecting surfaces perpendicular to z, we form a totally enclosed rectangular cavity. We now cannot use the solution we have so far written, involving propagation along z; we must rather superpose a reflected wave traveling along $-z$, giving a standing wave along z. To satisfy the boundary conditions at the mirrors, which may be located for example at $z = 0$, $z = M$, we must make the sinusoidal function along z, whose wave length is λ_g, satisfy a condition analogous to (1.5), or

$$M = \frac{m\lambda_g}{2},\qquad(1.9)$$

where m is an integer. Combining (1.5), (1.9), and (1.8), this leads to

$$\left(\frac{m}{2M}\right)^2 + \left(\frac{n}{2L}\right)^2 = \frac{1}{\lambda_0^2}.\qquad(1.10)$$

That is, the free-space wave length, or the frequency, of the disturbance in the cavity is determined by the integers n and m. We have a discrete set of normal modes of oscillation, in which the cavity can resonate.

The problem is similar to that of the elastic vibrations of a coupled system or of a vibrating membrane in mechanics, in which there are a discrete set of normal modes of oscillation, each with its own resonant frequency, and characteristic wave form. Furthermore, as in mechanics, we can introduce normal coordinates describing the various modes, using these normal coordinates to describe forced oscillations of the system, and we can prove orthogonality relations holding between the various normal vibrations. These subjects are, however, beyond the scope of the present book. We now proceed to a much more general discussion of propagation in wave guides, holding for a wave guide of arbitrary cross section, and reducing to the results of the present section in our present simple case. This general discussion is simpler and more powerful, but it does not exhibit the way in which the field is made up of the superposition of direct and reflected waves.

2. Electromagnetic Field in the Wave Guide.—We shall assume a wave guide in the form of a hollow pipe of arbitrary cross section in the xy plane, extending indefinitely along z. As shown in Eqs. (1.2), Chap. VIII, each of the components of **E** and **H** must satisfy the wave equation, which, in empty space, is

$$\nabla^2 \mathbf{E} - \frac{1}{c^2}\frac{\partial^2 \mathbf{E}}{\partial t^2} = 0, \qquad \nabla^2 \mathbf{H} - \frac{1}{c^2}\frac{\partial^2 \mathbf{H}}{\partial t^2} = 0. \qquad(2.1)$$

We now ask if we can obtain a solution of the form $u(x,y)e^{j[\omega t - (2\pi z/\lambda_g)]}$, where $u(x,y)$ is a function of x and y, and where λ_g, the guide wave length, has the same significance as in Sec. 1. Substituting in (2.1), and using $\lambda_0 = 2\pi c/\omega$, we find that u satisfies the equation

$$\frac{\partial^2 u}{\partial x^2} + \frac{\partial^2 u}{\partial y^2} + \left(\frac{2\pi}{\lambda_c}\right)^2 u = 0, \qquad(2.2)$$

where λ_c is defined in terms of λ_g and λ_0 by (1.8). Equation (2.2) must be solved subject to certain boundary conditions, which we shall discuss presently. Subject to these boundary conditions, we find solutions for only a discrete set of values λ_c. From (1.8) we are then led, as in the preceding section, to the conclusion that λ_c forms a cutoff wave length, such that free-space wave lengths greater than λ_c are attenuated in the mode in question, while shorter free-space wave lengths are propagated. Thus those conclusions are quite general, and are not limited to the case of propagation between parallel reflecting planes.

The fields **E** and **H** must satisfy not only the wave equation, but Maxwell's equations as well. When we write these equations down, we find at once, as mentioned in the preceding section, that two separate types of solutions are possible: solutions for which $E_z = 0$, the transverse electric, or *TE*, waves, and solutions for which $H_z = 0$, the transverse magnetic, or *TM*, waves. We shall let $\mathbf{E}_t$, $\mathbf{H}_t$ be the transverse components of **E** and **H** (that is, $\mathbf{E}_t = \mathbf{i}E_x + \mathbf{j}E_y$, and so on). Letting **k** be unit vector along the z axis, as usual, we then find easily and directly from Maxwell's equations the following relations:

$$TE: \text{grad } H_z = 2\pi j \frac{\lambda_g}{\lambda_c^2} \mathbf{H}_t,$$

$$TM: \text{grad } E_z = 2\pi j \frac{\lambda_g}{\lambda_c^2} \mathbf{E}_t. \tag{2.3}$$

In these expressions, H_z and E_z represent that part of the corresponding quantity which must be multiplied by the exponential $e^{j[\omega t - (2\pi z/\lambda_g)]}$ to give the complete components. Since they are functions of x and y only, their gradients are in the xy plane, as is proper for $\mathbf{H}_t$ or $\mathbf{E}_t$. We find that there is a relationship between the transverse components of **E** and **H,** as follows:

$$\mathbf{H}_t = \frac{\mathbf{k} \times \mathbf{E}_t}{Z_0}, \text{ where } Z_0 = \sqrt{\frac{\mu_0}{\epsilon_0}} \frac{\lambda_g}{\lambda_0} \text{ for } TE$$

$$= \sqrt{\frac{\mu_0}{\epsilon_0}} \frac{\lambda_0}{\lambda_g} \text{ for } TM. \tag{2.4}$$

This means, since **k** and $\mathbf{E}_t$ are at right angles to each other, that $\mathbf{H}_t$ is at right angles to $\mathbf{E}_t$ in the xy plane, and is equal in magnitude to the magnitude of $\mathbf{E}_t$, divided by the quantity Z_0, which is clearly analogous to the corresponding quantity introduced in Eq. (2.1), Chap. VIII.

Using (2.3) and (2.4), we can find all the components of **E** and **H** from H_z (in the TE case) or from E_z (in the TM case). These quantities, like the transverse components, satisfy the wave equation (2.2), and are scalar solutions of that differential equation. Furthermore, E_z is zero on the boundary of the guide (since **E** must have no tangential component on the surface), while H_z has a vanishing normal derivative on the boundary (since **H** must have no normal component on the surface, and $\mathbf{H}_t$ is proportional to the gradient of H_z). We may draw lines of H_z = constant (in the TE case) or of E_z = constant (in the TM case) in the xy plane. Then by (2.3) the orthogonal trajectories of these lines will be along the direction of $\mathbf{H}_t$ (in the TE case) or of $\mathbf{E}_t$ (in the TM case). Finally, since by (2.4) the direction of $\mathbf{H}_t$ is perpendicular to that of $\mathbf{E}_t$, the lines of constant H_z will be along the direction of $\mathbf{E}_t$ (in the TE case), and the lines of constant E_z will be along the direction of $\mathbf{H}_t$ (in the TM case). Proceeding in this way, we may draw lines of force, for the transverse components of **E** and **H,** in the xy plane, finding of course that the electric lines of force meet the surfaces at right angles, while the magnetic lines of force are tangential to the surface.

For every scalar solution of the two-dimensional wave equation satisfying the condition that it vanishes on the boundary, we get a TM wave, and for every solution whose normal derivative vanishes we get a TE wave, as we have seen above. There will be an infinite number of solutions of each type, each corresponding to a particular cutoff wave length. These wave lengths may be arranged in order of decreasing magnitude; they start with a largest cutoff wave length, associated with the lowest mode of oscillation, and extend indefinitely toward shorter and shorter wave lengths, so that we have an infinite number of modes of oscillation. We shall give in the next section some simple examples of the modes in various types of wave guides.

In some cases, the first mode has an infinite cutoff wave length; in this case we call it a "principal mode." When a mode of infinite cutoff wave length, or principal mode, exists, it has great practical importance, because it can be used to propagate any wave length, no matter how long. The commonly used mode of the coaxial line is a principal mode, and the familiar parallel-wire transmission line, ordinarily used for low frequencies, can also be considered as a wave guide with a principal mode. We find that such a principal mode exists only if the wall of the wave guide consists of at least two separated conductors, as for instance in the coaxial line. The physical reason for this is quite clear: we can put a very low frequency, or direct current, into a transmission line consisting of two or more conductors, and they will be insulated from each other, and suited to conduct the current. If there is only one conductor, however, as in an ordinary hollow pipe, there would clearly be a short circuit for a low frequency or direct current, and no propagation is possible until we get to a wave length short enough so that something like real wave propagation occurs. For a principal mode, the cutoff wave length is infinite, so that (2.3) tells us that grad H_z, or grad E_z, must be zero. That is, the longitudinal components of both **E** and **H** are zero, and such a wave is simultaneously transverse electric and transverse magnetic. It is sometimes called a "transverse electromagnetic (*TEM*) wave" for this reason. Furthermore for such a wave, as we see from (1.8), the guide wave length becomes equal to the free-space wave length, so that the velocity of propagation becomes c. Finally, from (2.4), the quantity Z_0 for a principal wave becomes $\sqrt{\mu_0/\epsilon_0}$, which by Eq. (2.1), Chap. VIII, is the ratio of magnitudes of **E** and **H** in a plane wave in free space. Thus a principal wave has many of the properties of a wave in free space.

3. Examples of Wave Guides.—The commonest type of wave guide is that of rectangular cross section, bounded by planes at $x = 0$, $x = a$, $y = 0$, $y = b$. The solution of the wave equation (2.2) in this case can be carried out in rectangular coordinates, and is at once obvious. Thus for a *TE* wave we have

$$H_z = \cos \frac{m\pi x}{a} \cos \frac{n\pi y}{b} \, e^{j[\omega t - (2\pi z/\lambda_g)]}, \tag{3.1}$$

where

$$\frac{1}{\lambda_c^2} = \left(\frac{m}{2a}\right)^2 + \left(\frac{n}{2b}\right)^2, \tag{3.2}$$

and where m, n are integers. Using (2.3), (2.4), we then have for th $_2$ other components

$$H_x = -\frac{\lambda_c^2}{2\pi j \lambda_g} \frac{m\pi}{a} \sin \frac{m\pi x}{a} \cos \frac{n\pi y}{b} \, e^{j(\omega t - 2\pi z/\lambda_g)},$$

$$H_y = -\frac{\lambda_c^2}{2\pi j \lambda_g} \frac{n\pi}{b} \cos \frac{m\pi x}{a} \sin \frac{n\pi y}{b} \, e^{j(\omega t - 2\pi z/\lambda_g)},$$

$$E_x = -\sqrt{\frac{\mu_0}{\epsilon_0}} \frac{\lambda_c^2}{2\pi j \lambda_0} \frac{n\pi}{b} \cos \frac{m\pi x}{a} \sin \frac{n\pi y}{b} \, e^{j(\omega t - 2\pi z/\lambda_g)},$$

$$E_y = \sqrt{\frac{\mu_0}{\epsilon_0}} \frac{\lambda_c^2}{2\pi j \lambda_0} \frac{m\pi}{a} \sin \frac{m\pi x}{a} \cos \frac{n\pi y}{b} \, e^{j(\omega t - 2\pi z/\lambda_g)},$$

$$E_z = 0. \tag{3.3}$$

Of this infinite set of possible *TE* solutions, only one is commonly used in practice, in a rectangular guide. If $a > b$, this is the mode of longest cutoff wave length, given by $m = 1, n = 0$. We note from (3.3) that for this mode $E_z = 0$, and E_y is proportional to $\sin \pi x/a$, going to a maximum at the center of the guide, and is independent of y, so that the lines of force run straight across the guide. There is no mode corresponding to $m = 0, n = 0$, for then as we see from (3.3) all the transverse components of field are zero, and we verify easily that the field cannot exist. Thus there is no principal mode for the rectangular guide.

For the TM waves, we have

$$E_z = \sin \frac{m\pi x}{a} \sin \frac{n\pi y}{b} e^{j(\omega t - 2\pi z/\lambda_g)}, \tag{3.4}$$

where the cutoff wave length is again given by (3.2). We thus see that, for the case of the rectangular guide, the TE and TM waves have the same cutoff wave lengths; this is a special case, which does not hold for most shapes of guides. Clearly we must have m and n both equal to or greater than unity, however, in the case of the TM waves, as we see from (3.4), so that the case $m = 1, n = 0$, which provides the dominant TE wave, does not exist for the TM waves. From (3.4) we can find the other components of **E** and **H**, as we did for the TE waves in (3.3).

The next most important example of wave guides after the rectangular guide is that in which the bounding surfaces are circular cylinders. This includes both the circular guide, and the coaxial line, which has two concentric cylinders, with the wave propagated in the annular space between. For either of these cases, we must handle Maxwell's equations and the wave equation in polar coordinates. In Appendix IV, where we discuss vector operations in curvilinear coordinates, we see that the wave equation, equivalent to (2.2), is

$$\frac{1}{r} \frac{\partial}{\partial r} \left(r \frac{\partial u}{\partial r} \right) + \frac{1}{r^2} \frac{\partial^2 u}{\partial \theta^2} + \left(\frac{2\pi}{\lambda_c} \right)^2 u = 0. \tag{3.5}$$

This equation can be solved by separation of variables. We assume that u can be written as a product of a function of r, and a function of θ: $u = R(r)\Theta(\theta)$. Substituting, dividing by u, and multiplying by r^2, (3.5) takes the form

$$\frac{r^2}{R} \left[\frac{1}{r} \frac{d}{dr} \left(r \frac{dR}{dr} \right) + \left(\frac{2\pi}{\lambda_c} \right)^2 R \right] + \frac{1}{\Theta} \frac{d^2\Theta}{d\theta^2} = 0. \tag{3.6}$$

The first term of (3.6) is a function of r alone, the second a function of θ alone; hence by the usual argument used in separation of variables, each term must be a constant. Let us assume then that

$$\frac{1}{\Theta} \frac{d^2\Theta}{d\theta^2} = -n^2, \qquad \frac{d^2\Theta}{d\theta^2} + n^2\Theta = 0, \qquad \Theta = \sin n\theta \text{ or } \cos n\theta. \quad (3.7)$$

Since u must be a single-valued function of position, it is clear that increasing θ by 2π must bring Θ back to its original value; thus the quantity n must be an integer. Substituting the value from (3.7) back in (3.6), and multiplying by R/r^2, we then have as the equation for R

$$\frac{1}{r} \frac{d}{dr}\left(r\frac{dR}{dr} \right) + \left[\left(\frac{2\pi}{\lambda_c}\right)^2 - \frac{n^2}{r^2} \right] R = 0. \qquad (3.8)$$

Equation (3.8) is Bessel's equation, whose properties are discussed in Appendix VII. It has two independent solutions, called Bessel's function and Neumann's function:

$$R = J_n\left(\frac{2\pi r}{\lambda_c}\right) \qquad \text{or} \qquad N_n\left(\frac{2\pi r}{\lambda_c}\right). \qquad (3.9)$$

These functions have the following properties, discussed in Appendix VII: At $x = 0$, $J_n(x)$ is proportional to x^n, whereas $N_n(x)$ becomes infinite. For large values of x, $J_n(x)$ and $N_n(x)$ approach the values

$$J_n(x) \rightarrow \sqrt{\frac{2}{\pi x}} \cos\left(x - \frac{2n+1}{4}\pi \right),$$
$$N_n(x) \rightarrow \sqrt{\frac{2}{\pi x}} \sin\left(x - \frac{2n+1}{4}\pi \right). \qquad (3.10)$$

From this, it is clear that, in a circular guide, where the field must be finite at $r = 0$, we must use only the Bessel functions J_n. On the other hand, with a coaxial line, in which $r = 0$ does not fall within the region where we are finding the field, the Neumann as well as the Bessel functions can, and in fact must, be used.

We see, then, that the function u which satisfies the wave equation (3.5) must be the form

$$u = \left[AJ_n\left(\frac{2\pi r}{\lambda_c}\right) \cos(n\theta - a) + BN_n\left(\frac{2\pi r}{\lambda_c}\right) \cos(n\theta - b) \right]. \quad (3.11)$$

Here we have introduced amplitude constants A and B, and phase constants a and b. For a *TE* wave, as we have seen before, u stands for H_z, and its normal derivative must be zero at the boundary of the guide; that is, the derivative of the function of r with respect to r

must be zero. For a TM wave, u stands for E_z, which must itself be zero on the boundary. Now let us consider first a circular wave guide, for which we need only the Bessel function J_n. If R_1 is the radius of the guide, we must clearly have $J_n(2\pi R_1/\lambda_c) = 0$ for a TM mode, and $J'_n(2\pi R_1/\lambda_c) = 0$ for a TE mode, where the prime indicates a derivative.

By giving a table of the maxima, minima, and zeros of the Bessel's functions, we can then find the values of λ_c for the various modes. The first few values are given in the following table:

MAXIMA AND ROOTS OF BESSEL'S FUNCTIONS

$J'_n(x_{nm}) = 0(TE)$	$x_{01} = 3.832$	$x_{11} = 1.842$	$x_{21} = 3.05$
	$x_{02} = 7.016$	$x_{12} = 5.330$	$x_{22} = 6.71$
$J_n(x_{nm}) = 0(TM)$	$x_{01} = 2.405$	$x_{11} = 3.832$	$x_{21} = 5.135$
	$x_{02} = 5.520$	$x_{12} = 7.016$	$x_{22} = 8.417$

If we let x_{nm} be the mth value of x for which $J'_n(x) = 0$ (for the TE modes) or for which $J_n(x) = 0$ (for the TM modes), we then clearly have the cutoff wave length given by

$$\lambda_c = \frac{2\pi R_1}{x_{nm}}, \tag{3.12}$$

where the x_{nm}'s are given in the table above. Thus we can calculate the cutoff wave lengths of the various modes. Using (3.11), and the relations (2.3) and (2.4), we can then find the various components of the field, where we remember that the gradient must be computed in polar coordinates, as discussed in Appendix IV.

For a coaxial line, we must satisfy boundary conditions, not at one value of r, but at two. This can be done only by combining the Bessel and Neumann functions, and using the extra arbitrary constant gained thereby. Thus, if the radius of the inner conductor is R_1, and of the outer one is R_2, we may combine the Bessel and Neumann functions in such a way as to satisfy the boundary conditions at R_1. From (3.10) it is clear that a suitable combination of these two functions will act like a cosine or sine function with arbitrary phase, which can be chosen to put the zero, or the maximum, at any desired point. This can be done for arbitrary λ_c. Then we can choose λ_c so as to satisfy the boundary condition at R_2 as well. This process is a little involved, and it is not so easy to get the numerical answers to problems as with the circular guide, where we found the cutoff wave lengths explicitly in (3.12). The main interest in the coaxial line, however, is not in the modes that we investigate in this way, but in the principal mode, whose cutoff wave length is infinite. This can be approached by a limiting process from the solution in terms of Bessel and Neumann functions, but it is much easier to discuss it directly from the wave equation (2.2).

For a principal mode, where λ_c is infinite, the equation (2.2) reduces to Laplace's equation $\partial^2 u/\partial x^2 + \partial^2 u/\partial y^2 = 0$. Thus the problem of finding u becomes mathematically equivalent to that of finding an electrostatic potential in a two-dimensional problem. In our case of circular symmetry, we may take u to be ln r, so that its gradient points in the direction of r, and is a constant divided by r. With a coaxial line, this gradient may be chosen as **E**, with **H** at right angles to it, and we have satisfied our boundary conditions at the surfaces of the conductors. Thus we have, for the principal mode of a coaxial line, the very simple solutions

$$E_r = \frac{1}{r} e^{j(\omega t - 2\pi z/\lambda_0)}, \qquad H_\theta = \sqrt{\frac{\epsilon_0}{\mu_0}} E_r. \qquad (3.13)$$

The wave, like all principal modes, is propagated with the velocity of light. The mode, as in all principal modes, is *TEM*, both **E** and **H** being transverse. This is not the only case of principal modes that we can work out simply. For instance, for a parallel-wire transmission line, we may use our earlier solution for the capacity of two parallel cylindrical conductors, worked out in Sec. 2, Chap. II. There we found the potential of two such conductors, a solution of Laplace's equation, reducing to a constant on the surface of each conductor. The gradient of this potential gave the electrostatic field, cutting each conductor at right angles. By the principles we have just stated, this same field is proportional to the value of **E** in the principal mode of the parallel-wire transmission line, and **H** is at right angles to it, so that the equipotentials are also the lines of magnetic force.

4. Standing Waves in Wave Guides.—We have so far considered a wave propagated along the z direction in an infinitely long wave guide, and varying according to an exponential $e^{j(\omega t - 2\pi z/\lambda_g)}$. Now, however, we consider the possibility that the wave may be reflected from an obstacle at the end, sending back a reflected wave. This reflecting obstacle may be a perfect reflector, such as a metallic wall; then the reflected wave will be equal in amplitude to the incident wave, and there will be standing waves in which there are nodes and antinodes, the amplitudes canceling and resulting in zero disturbance at the nodes, with addition of the amplitudes midway between. On the other hand, the obstacle may be only a partial reflector, allowing some radiation to penetrate it, reflecting the rest. In this case, the reflected wave will have less intensity than the incident wave, and the interference of the two will never completely cancel the disturbance.

If the amplitude of the transverse electric field in the reflected wave is r times as great as in the incident wave, where r is sometimes called the "reflection coefficient," the amplitude of the resulting disturbance will vary from something proportional to $1 + r$, where the two waves add, to something proportional to $1 - r$, where they oppose each other. The ratio of maximum to minimum E, or $(1 + r)/(1 - r)$, is called the "standing-wave ratio" in voltage. We see that it can go from unity, when $r = 0$, and there is no reflection, to infinity, when $r = 1$, and there is perfect reflection. The points along the guide where the resultant transverse electric field is a maximum are called "standing-wave maxima," and those where it is a minimum are "standing-wave minima." For an infinite standing-wave ratio the standing-wave minima become the nodes, and in any case they come a half wave length apart. It is clear that from a measurement of the magnitude of the standing-wave ratio we can find the reflection coefficient r, and from the positions of the standing-wave minima we can find the phase change on reflection.

There is one simple case in which we can calculate the reflection coefficient easily. That is the case in which part of the guide, say for positive z, is filled with a dielectric, while the rest, say for negative z, is empty. Then a wave approaching the surface of separation from negative z's will be partly reflected, partly transmitted, just as a wave approaching a surface of glass from air is partly reflected, as we have seen in Chap. X. In simple cases, we could use Fresnel's equations, as derived in that chapter, to find the reflection coefficient in the present case, but we shall instead proceed directly, showing that the results are in close analogy to those of Chap. X. First we must consider the nature of the field in the part of the guide filled with the dielectric. If we follow through the derivation of Sec. 2, we find that the only change introduced by filling the space with a dielectric is to change ϵ_0 to ϵ. This has the effect first of changing the velocity of propagation. Thus the free-space wave length λ_0 is equal to $1/n$ times the value for empty space, where n is the index of refraction. For a given cutoff wave length, (1.8) then shows that λ_g will be smaller than in free space, though not in the same ratio as λ_0. In Eq. (2.2), the wave equation for u, we find that there is no change introduced as a result of the dielectric. Thus the problem of finding cutoff wave lengths and satisfying boundary conditions at the boundary of the guide is as before. In (2.3), the only change in the process of determining the transverse from the longitudinal components of field comes in the changed value of λ_g, which changes the scale of the transverse components, but not the directions of the vectors. Similarly in (2.4) the only change in the relation between transverse $\mathbf{E}$ and $\mathbf{H}$ comes from the change in the quantity Z_0, which is different both because of the direct change of ϵ and because of the change in λ_g.

Now let us consider a direct and a reflected wave for negative z, and a transmitted wave for positive z. If $\mathbf{E}_t^0$ represents the tangential electric field in its dependence on x and y, we may write the tangential electric and magnetic fields in the first medium as

$$\mathbf{E}_t = \mathbf{E}_t^0[e^{j(\omega t - 2\pi z/\lambda_g)} + re^{j(\omega t + 2\pi z/\lambda_g)}]$$
$$\mathbf{H}_t = \left(\frac{\mathbf{k} \times \mathbf{E}_t^0}{Z_0}\right)[e^{j(\omega t - 2\pi z/\lambda_g)} - re^{j(\omega t + 2\pi z/\lambda_g)}]. \tag{4.1}$$

Here the minus sign in the expression for the transverse magnetic field in the reflected wave is necessary so that the Poynting vector of the reflected wave will point along $-z$. Now at $z = 0$, at the boundary of the second medium, we have from (4.1)

$$\mathbf{H}_t = \frac{\mathbf{k} \times \mathbf{E}_t}{Z_0}\frac{1 - r}{1 + r}. \tag{4.2}$$

But, because of the continuity of the tangential components of $\mathbf{E}$ and $\mathbf{H}$, the values of $\mathbf{H}_t$ and $\mathbf{E}_t$ in (4.2) are also the values found in the second medium. On the other hand, by the modification of (2.4) appropriate for the second medium, we must also have

$$\mathbf{H}_t = \frac{\mathbf{k} \times \mathbf{E}_t}{Z_0'}, \tag{4.3}$$

where Z_0' is the value for the second medium. Equating (4.2) and (4.3), we may solve for r, finding

$$r = \frac{Z_0' - Z_0}{Z_0' + Z_0}. \tag{4.4}$$

In (4.4) we have solved the problem of finding the reflection coefficient for the electric field, at a boundary between two media in a wave guide. The analogy of this to Eq. (3.4), Chap. X, is obvious, though the meaning of the symbols is somewhat different. Proceeding in the other direction, we have

$$\frac{1 + r}{1 - r} = \frac{Z_0'}{Z_0}. \tag{4.5}$$

Thus we see that a measurement of reflection coefficient allows us to find the quantity Z_0' characteristic of the second medium.

Often in actual practice there are other sorts of discontinuities in wave guides in addition to a simple change of index of refraction. For instance, there may be a diaphragm in the guide, with a hole of some arbitrary shape in it. At such an obstacle, part of the wave will be reflected, part transmitted. We can then set up a reflection coefficient, just as we have done in the case we have discussed, and can get the properties of the reflection by measuring the standing-wave ratio and the position of the standing-wave minima. In such cases it is often convenient to define a quantity like the Z_0' of (4.5), characterizing the obstacle. We have already pointed out, in Sec. 2, Chap. VIII, the analogy between quantities like Z_0 and an impedance; in this way we are allowed to assign certain impedances to obstacles in wave guides, which prove to be pure reactances in case they introduce no losses. A further discussion of such reactances is beyond the scope of this text.

5. Resonant Cavities.—We have already pointed out that, by terminating a guide with two perfectly reflecting walls, we form a resonant cavity. At each such wall, say at $z = 0$, $z = M$, the field will be perfectly reflected, so that there will be a standing-wave pattern set up within the cavity, with infinite standing-wave ratio. To satisfy boundary conditions at both walls, we must have the tangential electric field zero at each, and this can be accomplished only if there is an integral number of half wave lengths in the length M. Thus we have the condition

$$M = \frac{m\lambda_g}{2},\qquad (5.1)$$

where m is an integer. Putting this into Eq. (1.8), we then have

$$\frac{1}{\lambda_0^2} = \frac{1}{\lambda_c^2} + \left(\frac{m}{2M}\right)^2. \qquad (5.2)$$

That is, the free-space wave length, and frequency, are definitely determined for a cavity resonance. For each cutoff wave length of the guide, we shall have an infinite number of resonant modes of the resonant cavity; so that, since in general there will be a double infinity of cutoff wave lengths, there will be a triple infinity of resonant modes for the cavity. We can, of course, get formulas for the resonant frequencies in any case in which we can compute the cutoff wave lengths, as in the cases of rectangular and circular guides which we have already discussed.

The resonant cavities formed in this way, by putting reflecting walls at the ends of a guide, form but a special case of the general problem of resonant cavities. Resonant modes of oscillation can exist in any cavity bounded by perfectly reflecting walls, no matter what its shape. In some cases we can solve for the fields (as for instance in the spherical cavity, which we shall mention in the next chapter), but in the great majority of cases there is no practical way to solve the wave equation. We cannot in general reduce the solution of Maxwell's equations to the solution of a scalar wave equation like (2.2), as we have been able to do for the guide; we must find vector solutions from the outset. The problem of discussing fields inside an arbitrary resonant cavity is then a complicated one, but nevertheless we can prove certain general theorems about them, even without being able to solve for them analytically. These theorems are, however, too complicated for us to take up here.

Problems

1. A wave travels along an air-filled coaxial line having an inner conductor of radius a and a sheath of inner radius b, both of negligible resistance, the peak voltage and current of this wave being V and I. Using cylindrical coordinates, with the z axis coincident with the axis of the line, show that the electric and magnetic fields are given by

$$E_r = \frac{V e^{i(\omega t - \beta z)}}{r \ln b/a}, \qquad H_\theta = \frac{I e^{i(\omega t - \beta z)}}{2\pi r}$$

What value must β have? Find the characteristic impedance V/I and the wave impedance E_r/H_θ in terms of ϵ_0, μ_0, a, and b.

2. Starting with Eq. (3.4) for the longitudinal electric-field component, derive expressions for the transverse components of **E** and **H** for TM waves in a rectangular wave guide.

3. For the case of TE and TM modes in a rectangular wave guide, compute Poynting's vector, its average value, and from this the power transmitted by these modes. For the lowest TE modes in a rectangular wave guide, find an expression for the maximum power that can be transmitted if the electric field cannot exceed a value E_0 without sparking.

4. Find expressions for the field components for TE and TM waves in a circular wave guide. Discuss the degeneracies present in these modes, especially those resulting from the relation $J_0'(x) = -J_1(x)$.

5. Show by direct substitution in Maxwell's equations that the following field for a TM wave is a solution of these equations:

$$H_x = -\frac{k^2}{\beta\omega\mu} E_y = H_0 e^{\pm j\sqrt{k^2-\beta^2}\,y} e^{i(\omega t - \beta z)}$$

$$E_z = \mp \frac{\omega\mu\sqrt{k^2-\beta^2}}{k^2} H_x,$$

where $k^2 = \omega^2\epsilon\mu - j\omega\mu\sigma$. Use this solution to discuss the propagation of a surface wave along a metal surface.   Let the metal fill the space $y < 0$, with air for $y > 0$. Write the appropriate solutions in both regions, and show from the boundary conditions that β is given by $1/\beta^2 = 1/k^2 + 1/k_0^2$, with $k_0^2 = \omega^2 \epsilon \mu$.   Now assume that the conductivity σ of the metal is high so that $|k^2| \gg k_0^2$.   Compute the total current in the metal per unit width in the x direction, and show that it is equal to the magnetic field at the metal-air surface, just as if one had a perfect conductor. From the y component of the Poynting vector, compute the power flow into the metal at the boundary $y = 0$, and show that it is of the form $I^2 R$, with I the rms total current in the metal.   Show further that this resistance R is the same as if the current density were uniform to a depth below the metal surface equal to the skin depth $\sqrt{2/\omega\mu\sigma}$ and zero below this.

6. A wave guide is terminated by a thin conducting surface perpendicular to the axis of the guide.   Find the resistance per square of this conductor (that is, the resistance across a square of the material, from one side to the opposite side, which is independent of the size of the square), such that a wave falling on the surface will be totally absorbed, with no reflection.   Find the relation of this resistance per square to the quantity Z_0 of (2.4).

7. A wave guide is filled with a dielectric giving a value of Z_0 [from (2.4)] for $z < 0$; another dielectric giving Z_1 for $0 < z < L$; and a third giving Z_2 for $L < z$. A wave travels in the direction of increasing z, and in addition there are reflected waves, in the direction of decreasing z, for $z < 0$, and for $0 < z < L$.   By considering the boundary conditions at the surfaces of separation, find the amplitude of the reflected wave for $z < 0$.   Show that this amplitude is zero if L equals a quarter of a guide wave length (in the material filling the section $0 < z < L$), and if $Z_1 = \sqrt{Z_0 Z_2}$.

CHAPTER XII

SPHERICAL ELECTROMAGNETIC WAVES

Suppose we have an electric charge oscillating sinusoidally with the time. This charge will send out a spherical electromagnetic wave, radiating in all directions. Several physical problems are connected with such a wave. First, the phenomenon may be on a large scale, as in a radio antenna. Radiation from a vertical antenna, as a matter of fact, can be treated approximately by replacing the antenna by such an oscillating charge. But also on a smaller scale we can treat the radiation of short electromagnetic waves, or in other words light, from an atom that contains oscillating electrons. The electrons may have been set in motion by heat or bombardment, in which case we have the treatment of the emission of light from a luminous body; or they may be in forced motion under the action of another light wave, as in the case of the scattering of light. As a first step in the discussion of these problems, we consider the solution of Maxwell's equations in spherical coordinates. This solution, similar to the one we found in Chap. III for the static problem, will reduce in a simple case to the field of an oscillating dipole, which we require for the applications mentioned above. At the same time, it will give us the field of any oscillating multipole. Other forms of the same solution will lead to the field inside a spherical resonant cavity, or more generally in a cavity resonator bounded by surfaces $r = $ constant, $\theta = $ constant, and $\varphi = $ constant, in a system of spherical polar coordinates. We pass then to the solution of Maxwell's equations, later taking up their applications.

1. Maxwell's Equations in Spherical Coordinates.—In spherical polar coordinates, r, θ, φ, we can define a vector such as $\mathbf{E}$ in terms of its components E_r, E_θ, E_φ, along the directions in which these coordinates increase. Then, using the methods of vector operations in curvilinear coordinates, which are discussed in Appendix IV, Maxwell's equations for free space, in the absence of charge and current, become

$$\frac{1}{r \sin \theta} \left[\frac{\partial}{\partial \theta} (\sin \theta \, E_\varphi) - \frac{\partial E_\theta}{\partial \varphi} \right] + \mu_0 \frac{\partial H_r}{\partial t} = 0$$

$$\frac{1}{r \sin \theta} \frac{\partial E_r}{\partial \varphi} - \frac{1}{r} \frac{\partial (r E_\varphi)}{\partial r} + \mu_0 \frac{\partial H_\theta}{\partial t} = 0$$

$$\frac{1}{r} \frac{\partial(rE_\theta)}{\partial r} - \frac{1}{r} \frac{\partial E_r}{\partial \theta} + \mu_0 \frac{\partial H_\varphi}{\partial t} = 0$$

$$\frac{1}{r^2} \frac{\partial}{\partial r} (r^2 H_r) + \frac{1}{r \sin \theta} \frac{\partial}{\partial \theta} (\sin \theta \, H_\theta) + \frac{1}{r \sin \theta} \frac{\partial H_\varphi}{\partial \varphi} = 0$$

$$\frac{1}{r \sin \theta} \left[\frac{\partial}{\partial \theta} (\sin \theta \, H_\varphi) - \frac{\partial H_\theta}{\partial \varphi} \right] - \epsilon_0 \frac{\partial E_r}{\partial t} = 0$$

$$\frac{1}{r \sin \theta} \frac{\partial H_r}{\partial \varphi} - \frac{1}{r} \frac{\partial(rH_\varphi)}{\partial r} - \epsilon_0 \frac{\partial E_\theta}{\partial t} = 0$$

$$\frac{1}{r} \frac{\partial}{\partial r} (rH_\theta) - \frac{1}{r} \frac{\partial H_r}{\partial \theta} - \epsilon_0 \frac{\partial E_\varphi}{\partial t} = 0$$

$$\frac{1}{r^2} \frac{\partial}{\partial r} (r^2 E_r) + \frac{1}{r \sin \theta} \frac{\partial}{\partial \theta} (\sin \theta \, E_\theta) + \frac{1}{r \sin \theta} \frac{\partial E_\varphi}{\partial \varphi} = 0. \quad (1.1)$$

The first three of these represent the equation curl $\mathbf{E} + \mu_0 \, \partial\mathbf{H}/\partial t = 0$, the fourth is div $\mathbf{H} = 0$, the next three are curl $\mathbf{H} - \epsilon_0 \, \partial\mathbf{E}/\partial t = 0$, and the last is div $\mathbf{E} = 0$, where as before the divergence equations contribute no new information not included in the others. The solutions of these equations represent, in general, waves propagated outward, or inward, along the radius. Thus the r direction can be considered the direction of propagation. As with the case of the wave guides considered in Chap. XI, there are two types of waves: transverse electric, or TE, in which the longitudinal component of the electric field, E_r, is zero; and transverse magnetic, TM, in which H_r is zero. We shall set up the solutions of both types, and consider their properties.

The solution of the wave-guide problem was simplified a great deal because we could assume from the outset that each component of the field varied exponentially along the z axis, the direction of propagation. It cannot be assumed in a corresponding way here that the components vary exponentially with r; as a matter of fact, their dependence on r is given by certain Bessel's functions, which approach an exponential form only for large values of r. Therefore we cannot at once carry through derivations like those of Eqs. (2.2), (2.3), and (2.4) of Chap. XI, expressing all the field components in the TE case algebraically in terms of H_z, and all the field components in the TM case in terms of E_z. Nevertheless an essentially equivalent discussion can be carried through, except that here the relations are not simply algebraic but involve differentiation.

Thus let us start with the TE case. We let H_r be a scalar function of r, θ, φ. It is slightly more convenient to operate, not with H_r, but with a quantity that we may denote u, defined by

$$u = rH_r. \quad (1.2)$$

It will appear that the equation that must be satisfied by u is the wave equation, which in spherical polar coordinates becomes

$$\frac{1}{r^2}\frac{\partial}{\partial r}\left(r^2\frac{\partial u}{\partial r}\right) + \frac{1}{r^2\sin\theta}\frac{\partial}{\partial\theta}\left(\sin\theta\frac{\partial u}{\partial\theta}\right)$$
$$+ \frac{1}{r^2\sin^2\theta}\frac{\partial^2 u}{\partial\varphi^2} - \frac{1}{c^2}\frac{\partial^2 u}{\partial t^2} = 0. \quad (1.3)$$

We assume a sinusoidal dependence on time, varying according to the exponential $e^{j\omega t}$. The equation can then be solved by separation of variables, just as in Sec. 2, Chap. III, where we discussed the corresponding static problem. Letting u be of the form

$$u = R(r)\Theta(\theta)\Phi(\varphi)e^{j\omega t}, \quad (1.4)$$

we find that the functions satisfy the equations

$$\frac{1}{r^2}\frac{d}{dr}\left(r^2\frac{dR}{dr}\right) + \left[\frac{\omega^2}{c^2} - \frac{l(l+1)}{r^2}\right]R = 0, \quad (1.5)$$

$$\frac{1}{\sin\theta}\frac{d}{d\theta}\left(\sin\theta\frac{d\Theta}{d\theta}\right) + \left[l(l+1) - \frac{m^2}{\sin^2\theta}\right]\Theta = 0, \quad (1.6)$$

$$\frac{d^2\Phi}{d\varphi^2} + m^2\Phi = 0, \quad (1.7)$$

similar to Eqs. (2.3), (2.4), and (2.1), of Chap. III. We shall discuss the solutions of these equations in the next section. Assuming that we have found u from them, we can now set up values for all the other field components.

It is not easy to proceed straightforwardly to a solution of Maxwell's equations. We find, however, the following equations for the field components in terms of u:

$$H_r = \frac{u}{r}$$

$$H_\theta = \frac{1}{l(l+1)}\frac{1}{r}\frac{\partial}{\partial\theta}\frac{\partial}{\partial r}(ru)$$

$$H_\varphi = \frac{1}{l(l+1)}\frac{1}{r\sin\theta}\frac{\partial}{\partial\varphi}\frac{\partial}{\partial r}(ru)$$

$$E_r = 0$$

$$E_\theta = \frac{-j\omega\mu_0}{l(l+1)}\frac{1}{r\sin\theta}\frac{\partial}{\partial\varphi}(ru)$$

$$E_\varphi = \frac{j\omega\mu_0}{l(l+1)}\frac{1}{r}\frac{\partial}{\partial\theta}(ru). \quad (1.8)$$

By direct substitution in Maxwell's equations, (1.1), we can show that these equations are satisfied if u obeys the wave equation (1.3), or its equivalent equations (1.5), (1.6), (1.7). Thus we prove that the functions (1.8) are the solutions of the problem which we desire. We note that, instead of defining H_θ and H_φ directly in terms of u, we can instead set up the relations

$$H_\theta = \frac{1}{j\omega\mu_0} \frac{1}{r} \frac{\partial}{\partial r} (rE_\varphi),$$
$$H_\varphi = \frac{-1}{j\omega\mu_0} \frac{1}{r} \frac{\partial}{\partial r} (rE_\theta). \tag{1.9}$$

We shall soon see that these equations form the analogy to the relation (2.4) of Chap. XI, giving the relation between the transverse components of **E** and **H** in a wave guide.

The relations for the transverse magnetic, *TM*, case are analogous to those just set up. In that case, we have a longitudinal component E_r of electric field, which we write as

$$E_r = \frac{v}{r}, \tag{1.10}$$

where v satisfies the same wave equation (1.3) that u satisfied. In terms of v, we can then write the other components as

$$E_\theta = \frac{1}{l(l+1)} \frac{1}{r} \frac{\partial}{\partial \theta} \frac{\partial}{\partial r} (rv)$$
$$E_\varphi = \frac{1}{l(l+1)} \frac{1}{r \sin \theta} \frac{\partial}{\partial \varphi} \frac{\partial}{\partial r} (rv)$$
$$H_r = 0$$
$$H_\theta = \frac{j\omega\epsilon_0}{l(l+1)} \frac{1}{r \sin \theta} \frac{\partial}{\partial \varphi} (rv)$$
$$H_\varphi = \frac{-j\omega\epsilon_0}{l(l+1)} \frac{1}{r} \frac{\partial}{\partial \theta} (rv) \tag{1.11}$$

where we have the relations

$$E_\theta = \frac{-1}{j\omega\epsilon_0} \frac{1}{r} \frac{\partial}{\partial r} (rH_\varphi)$$
$$E_\varphi = \frac{1}{j\omega\epsilon_0} \frac{1}{r} \frac{\partial}{\partial r} (rH_\theta). \tag{1.12}$$

We now have our complete set of equations for determining the fields, and shall consider their solutions in the next section.

2. Solutions of Maxwell's Equations in Spherical Coordinates.— To find the components of a field in spherical coordinates, we have seen that we must first find a scalar u or v, satisfying the wave equation (1.3), which separates to give (1.5), (1.6), and (1.7). The dependence on angle, given in (1.6) and (1.7), is just as it is in the static case taken up in Chap. III, and as in that case we find that $\Theta = P_l^m(\cos \theta)$, $\Phi = \sin m\varphi$ or $\cos m\varphi$. The equation for the function R, however, is more complicated than in the static case, though it reduces to it for $\omega = 0$. If we make the substitution $R = f/\sqrt{r}$, where f is a function of r, we find that (1.5) is transformed to Bessel's equation for a Bessel function of order $l + \frac{1}{2}$. Thus we have $R = J_{l+1/2}(\omega r/c)/\sqrt{r}$ or $N_{l+1/2}(\omega r/c)/\sqrt{r}$.

It is customary to define spherical Bessel and Neumann functions $j_l(x)$ and $n_l(x)$ by the equations

$$j_l(x) = \sqrt{\frac{\pi}{2x}}\, J_{l+1/2}(x), \qquad n_l(x) = \sqrt{\frac{\pi}{2x}}\, N_{l+1/2}(x). \qquad (2.1)$$

If l is an integer, which it will be in our applications, we can show, as is mentioned in Appendix VII, that j_l and n_l can be expressed in analytic form in terms of algebraic and trigonometric functions. For the first few functions we have

$$j_0(x) = \frac{\sin x}{x}, \qquad n_0(x) = -\frac{\cos x}{x}$$

$$j_1(x) = \frac{\sin x}{x^2} - \frac{\cos x}{x}, \qquad n_1(x) = -\frac{\sin x}{x} - \frac{\cos x}{x^2}. \qquad (2.2)$$

At large values of x, the terms in $1/x$ are the leading ones. In the general case, this leading term is given by

$$j_l(x) \to \frac{1}{x} \cos\left(x - \frac{l+1}{2}\pi\right),$$

$$n_l(x) \to \frac{1}{x} \sin\left(x - \frac{l+1}{2}\pi\right). \qquad (2.3)$$

In the opposite limit, as x tends to zero, we can expand in power series. In this case the leading term is

$$j_l(x) \to \frac{x^l}{1 \cdot 3 \cdot 5 \cdots (2l+1)}$$

$$n_l(x) \to -\frac{1 \cdot 1 \cdot 3 \cdot 5 \cdots (2l-1)}{x^{l+1}}. \qquad (2.4)$$

In terms of these spherical Bessel functions, it is clear that our functions u or v can be written in the form

$$u \text{ or } v = \left[j_l\left(\frac{\omega r}{c}\right) \text{ or } n_l\left(\frac{\omega r}{c}\right)\right] P_l^m(\cos \theta)(\sin m\varphi \text{ or } \cos m\varphi). \qquad (2.5)$$

The question of which functions to use in (2.5) depends on the boundary conditions to be satisfied, and this in turn depends on the problem that we are trying to solve. The simplest problem is the field within a cavity resonator bounded by a perfectly conducting spherical shell. In this case the field must be finite at the origin; therefore we can use only the functions j_l, not the Neumann functions n_l. Furthermore, the electric vector may have only an r component at the value of r giving the radius of the sphere. This means, for a *TE* mode, as we see from (1.8), that $j_l(\omega r/c)$ must be zero at the radius of the sphere, and for a *TM* mode, from (1.11), that the r derivative of $rj_l(\omega r/c)$ must be zero at this radius. These conditions easily determine the resonant frequencies of the various modes. We can almost as easily satisfy the boundary conditions for a resonator formed by the space between two concentric spheres. By making suitable combinations of $j_l(\omega r/c)$ and $n_l(\omega r/c)$, we see at once, from (2.2), that we can adjust the phase of the sinusoidal part of the function to any desired value. We then may choose the phase correctly to satisfy the boundary condition on one of the surfaces, and determine the frequency to satisfy the conditions on the other surface. These problems have an obvious analogy to the two-dimensional problems of the modes within a circular cylinder, and a coaxial line, discussed in the preceding chapter.

More interesting problems come when we take up the field in a region extending to infinity. In such cases, we are generally interested, not in standing waves, as in the solutions we have set up so far, but in traveling waves, traveling either inward or outward along r. Thus for instance in a problem of the emission of radiation from an atom or a radio antenna, we wish only waves traveling outward. It is easy to combine the Bessel and Neumann functions to secure traveling waves. The functions $j_l(x) \pm jn_l(x)$ [where we are to distinguish $j = \sqrt{-1}$ from $j_l(x)$] are called "spherical Hankel functions." From (2.3) we see that at large distances they approach the values

$$j_l(x) \pm jn_l(x) \rightarrow \frac{1}{x} e^{\pm j[x-(l+1)\pi/2]}. \tag{2.6}$$

Thus, if x is replaced by $\omega r/c$, we see that for the $+$ sign this represents the space part of a wave traveling in along r, and for the $-$ sign a wave traveling out, in each case with an amplitude that falls off inversely as the distance, so that the intensity, which is proportional to the square of the amplitude, falls off inversely as the square of the distance. For radiation from a source, then, we choose the $-$ sign in (2.6), and thus secure only an outgoing wave, without an incoming wave. We note that the Hankel functions, containing the Neumann functions, all have singularities at the origin. Thus an outgoing wave must either be propagated from a finite charge distribution, within which, since the charge density is different from zero, Maxwell's equations in the form (1.1) do not apply; in this case, a separate solution for the field within the charge distribution must be found, and it must be joined to the solution we have found; or the outgoing wave must be produced by a singularity of field, which can take the form of certain point charges, dipoles, and other multipoles. We shall take up examples of such singularities in a later section.

A very important form of problem is that of scattering of radiation by an object of some sort. If the object has spherical symmetry, as for instance if it is a small sphere of metal or of a dielectric, we can solve the problem of scattering by means of our spherical solutions of Maxwell's equations. For instance, let us consider scattering by a perfectly conducting sphere. Physically, we consider a plane wave incident on the sphere, and scattered spherical waves emerging from it. In this simple case, there is no mechanism for loss of energy, and all the energy scattered is made up from energy that strikes the sphere. We must make up the field by superposing solutions of the type we have considered. These solutions must satisfy boundary conditions of two sorts. First, at the surface of the sphere, the tangential components of electric field, and radial components of magnetic field, must be zero. To satisfy these, we must superpose Bessel and Neumann functions, in such a way as to make the function u equal to zero at the surface of the sphere for a *TE* mode, or the normal derivative of rv equal to zero for a *TM* mode. This can be done straightforwardly, and the result in general will not turn out to be a Hankel function, and hence will represent both an incoming and an outgoing spherical wave.

In addition to the boundary condition at the surface of the reflecting sphere, we must satisfy another condition, which requires more thought. We can easily state it physically: the disturbance must consist entirely of an incoming plane wave, plus outgoing spherical waves. To apply this condition, we must know how to describe a plane wave in terms of Bessel and Neumann functions. This is a rather involved problem, but in the similar case of a scalar wave equation, such as we should find for the scattering of sound, where we need only consider the pressure as a function of position, it becomes fairly simple. Considering only a scalar solution of the wave equation, one can prove the following theorem:

$$e^{\pm jkr \cos \theta} = \sum_{l=0}^{\infty} (2l + 1)(\pm j)^l P_l(\cos \theta) j_l(kr). \qquad (2.7)$$

Here $P_l(\cos \theta)$ is the value which $P_l^m(\cos \theta)$ takes on when $m = 0$. We note that $r \cos \theta = z$, so that, if we set $k = \omega/c$, the function on the left of (2.7) represents the space-dependent function which, combined with a time exponential $e^{j\omega t}$, would represent a plane wave traveling along the $\mp z$ direction. We thus see how to expand such a plane wave in terms of spherical Bessel functions. It is natural that no Neumann functions come into this expansion, for they would introduce a singularity at the origin, which the plane wave does not have.

To satisfy our requirements, we must then build up a solution of our problem, consisting of the sum of the plane wave (2.7), and only outgoing spherical waves. That is, the term of the solution associated with a given l value, and with $m = 0$ (for, as we easily see, this problem will involve only terms for $m = 0$), will have the form

$$(2l + 1)(\pm j)^l P_l(\cos \theta)\{j_l(kr) + \alpha_l[j_l(kr) - jn_l(kr)]\} \qquad (2.8)$$

where α_l is a coefficient measuring the amplitude of the lth scattered-wave component. Thus we see that the ratio of the coefficients of $j_l(kr)$ and $n_l(kr)$ must be $(\alpha_l + 1)/(-j\alpha_l)$. On the other hand, this ratio has been determined quite independently from the condition that the solution satisfy the boundary condition at the surface of the reflecting sphere. Thus, by equating these two values of the ratio of coefficients, we can evaluate the α_l's, and find the intensity of the scattered radiation.

This procedure in principle is simple, but in practice a complete discussion of the problem is beyond the scope of this text. We recall that our field is really a vector field, so that instead of (2.7) we must use a corresponding vector formulation of a plane wave in terms of spherical waves. When we do this, and carry through the discussion, we find the following sort of result: The behavior of the scattered radiation depends a great deal on the ratio of the diameter of the scattering sphere to the wave length. For a sphere very small compared with the wave length, the sphere acts almost as if it were in a uniform field. It then acquires a dipole moment, as we have shown in Chap. III, Sec. 4. This dipole alone scatters radiation, and we shall give an elementary discussion of this case in a later section of the present chapter, showing the type of radiation emitted from a dipole. As the sphere gets larger, it scatters a more and more complicated field, consisting now not only of dipole radiation, but of the fields of higher multipoles, which are discussed in Appendix VI. As the sphere becomes very large compared with a wave length, the outgoing wave takes on less and less of the character of a scattered wave, and approaches a reflected wave as predicted by geometrical optics for the simple optical problem of the reflection of a plane wave by a spherical mirror. In all these cases, we must consider not merely that part of the spherical wave which travels out in all directions, but in particular that part which travels in almost the same direction as the plane wave, after it has passed over the scattering obstacle. In this case, the scattered and incident waves interfere with each other, and it can be shown that they interfere destructively, so as to remove energy from the incident wave, just enough to supply the energy radiated out in the scattered wave. If the obstacle is small compared with the wave length, this interference fills a poorly defined region behind the obstacle, whose dimensions are determined by interference theory, but as the obstacle becomes large, the region of destructive interference becomes sharply defined, and becomes the ordinary shadow of geometrical optics. Around the edge of this shadow, however, there are diffraction fringes, alternations of intensity, resulting directly from the interference phenomenon.

All this interference and diffraction comes as part of a complete discussion of the problem of scattering of a plane wave by a sphere, and makes it clear why a complete treatment of the problem is beyond the scope of the present book. However, we shall handle interference and diffraction in an elementary way in a later chapter, bringing out many of the features that are present in the exact solution. The type of discussion sketched in the present section is important not only in electromagnetic theory, but in all other forms of wave theory as well. A simple example is the scattering of sound, as treated for example by P.M. Morse, *Vibration and Sound*, 2d ed. (McGraw-Hill Book Company, Inc., New York). More far-reaching is the application to quantum mechanics. The problem of scattering of an electron, neutron, or other colliding particle by an atom or nucleus can be handled to a first approximation, in wave mechanics, by a problem of scattering in a spherically symmetrical region where the index of refraction is a function of r. The mathematical framework of this problem is identical with that which we have sketched here, and a thorough understanding of the solution of the wave equation in spherical coordinates is an essential for the understanding of any type of scattering of waves by small obstacles.

3. The Field of an Oscillating Dipole.—In many ways the most important spherical wave physically is that produced by an oscillating electric dipole. This is the simplest *TM* wave. If we consider our solution (1.11), we see that the function v must depend on θ or φ, in order that any of the components of field may be different from zero. Thus we cannot have $l = 0$, since the spherical harmonic $P_0(\cos \theta)$ is a constant and does not depend on angles at all. The lowest value that l may have is 1. In this case, m can be 0 or 1. We consider the case $l = 1$, $m = 0$. This proves to be the simple dipole field. If we choose v to be

$$v = \cos \theta [j_1(kr) - jn_1(kr)],$$

or

$$v = \cos \theta \left[-\frac{1}{kr} + \frac{j}{(kr)^2} \right] e^{-jkr}, \tag{3.1}$$

where $k = \omega/c$, which is equivalent to it according to (2.2), we find for the field components

$$E_r = ke^{j(\omega t - kr)} \cos \theta \left[-\frac{1}{(kr)^2} + \frac{j}{(kr)^3} \right]$$

$$E_\theta = \frac{k}{2} e^{j(\omega t - kr)} \sin \theta \left[-\frac{j}{kr} - \frac{1}{(kr)^2} + \frac{j}{(kr)^3} \right]$$

$$H_\varphi = \frac{k}{2} \sqrt{\frac{\epsilon_0}{\mu_0}} e^{j(\omega t - kr)} \sin \theta \left[-\frac{j}{kr} - \frac{1}{(kr)^2} \right]. \tag{3.2}$$

The other field components are zero. These functions, of course, are to be multiplied by an arbitrary amplitude.

To understand the meaning of our solution, let us first consider the terms that are important at small distances, those in the highest inverse powers of r ($1/r^3$ for E_r and E_θ, $1/r^2$ for H_φ). We shall show that the terms in the electric force represent the field of an electric dipole at the origin and that the magnetic field is the field of the corresponding current element, derived from the time rate of change of the dipole moment, as found from the Biot-Savart law. We remember that a dipole of moment M has a potential

$$\psi = \frac{(M \cos \theta)}{4\pi\epsilon_0 r^2},$$

as we saw in Eq. (5.1), Chap. III. The field is then

$$E_r = -\frac{\partial \psi}{\partial r} = \frac{2M}{4\pi\epsilon_0} e^{j\omega t} \frac{\cos \theta}{r^3},$$

$$E_\theta = -\frac{1}{r} \frac{\partial \psi}{\partial \theta} = \frac{M}{4\pi\epsilon_0} e^{j\omega t} \frac{\sin \theta}{r^3}. \tag{3.3}$$

We observe that these are just a constant, $-jk^2M/2\pi\epsilon_0$, times the corresponding terms in (3.2). Similarly we note that a dipole whose moment was $Me^{j\omega t}$ would carry with it a current element equal to its time derivative, or to $j\omega Me^{j\omega t}$. This small current element would have a magnetic field H_φ, which would be given according to the Biot-Savart law, as described in Sec. 1, Chap. V, by

$$H_\varphi = \frac{j\omega M}{4\pi} e^{j\omega t} \frac{\sin\theta}{r^2}. \tag{3.4}$$

This is likewise $-jk^2M/2\pi\epsilon_0$ times the value from (3.2), at small distances. It is clear, then, that if we multiply (3.2) by this factor, we shall find a field that is a correct solution of Maxwell's equations at all distances, and yet that reduces to the field of a dipole of moment $Me^{j\omega t}$ at small distances, so that it must represent correctly the field of such a dipole. The field is

$$E_r = \frac{Mk^3}{4\pi\epsilon_0} e^{j(\omega t - kr)} \cos\theta \left[\frac{2j}{(kr)^2} + \frac{2}{(kr)^3} \right]$$

$$E_\theta = \frac{Mk^3}{4\pi\epsilon_0} e^{j(\omega t - kr)} \sin\theta \left[-\frac{1}{kr} + \frac{j}{(kr)^2} + \frac{1}{(kr)^3} \right]$$

$$H_\varphi = \frac{j\omega Mk^2}{4\pi} e^{j(\omega t - kr)} \sin\theta \left[\frac{j}{kr} + \frac{1}{(kr)^2} \right]. \tag{3.5}$$

4. The Field of a Dipole at Large Distances.—We have seen that at sufficiently small distances the field of an oscillating dipole reduces to what we should compute by Coulomb's law and the Biot-Savart law. As kr becomes appreciable compared with unity (that is, as $2\pi r/\lambda$ becomes appreciable, or as r approaches $1/2\pi$ wave lengths) the remaining terms become important. This is an example of a general situation: at distances from an oscillating charge distribution that are small compared with wave length, we can use Coulomb's law and the Biot-Savart law. On the other hand, for distances large compared with a wave length, the situation is entirely reversed, and the first term in each expression in (3.5) becomes the important one. The leading terms in E_θ and H_φ vary as $1/r$, while the leading term of E_r goes as $1/r^2$, and therefore can be neglected at sufficiently large distances. The field, in other words, becomes transverse at large distances, with an amplitude inversely proportional to the distance. In this range it is ordinarily referred to as the "radiation field." We see that **E** and **H** become at right angles to each other, and check further that the ratio of E_θ to H_φ approaches the value $\sqrt{\mu_0/\epsilon_0}$, characteristic of plane waves in free space.

The intensity varies as $\sin^2 \theta$, so that the maximum radiation is at right angles to the axis of the dipole. To measure the intensity, we may compute the average value of Poynting's vector. We have

$$
\begin{aligned}
S_r &= \tfrac{1}{2} Re E_\theta \bar{H}_\varphi \\
&= \frac{1}{2} Re \frac{\omega M^2 k^5}{16\pi^2 \epsilon_0} \sin^2 \theta \left[\frac{1}{(kr)^2} - \frac{j}{(kr)^5} \right] \\
&= \frac{\mu_0 \sqrt{\epsilon_0 \mu_0}\, \omega^4 M^2 \sin^2 \theta}{32\pi^2 r^2}.
\end{aligned}
\tag{4.1}
$$

We see that the intensity of radiation varies as $1/r^2$, the inverse-square law, as of course it must if the total flux outward through a sphere is independent of the size of the sphere. To find this total flux, we must integrate over all directions, by multiplying by the element of area $2\pi r^2 \sin \theta\, d\theta$ and integrating from 0 to π. We then have

$$
\begin{aligned}
\int S\, da &= \frac{\mu_0 \sqrt{\epsilon_0 \mu_0}\, \omega^4 M^2}{32\pi^2} 2\pi \int_0^\pi \sin^3 \theta\, d\theta \\
&= \frac{\mu_0 \sqrt{\epsilon_0 \mu_0}\, \omega^4 M^2}{12\pi}.
\end{aligned}
\tag{4.2}
$$

This is a well-known formula for the radiation from a dipole. The two essential features are that the radiation is proportional to the square of the amplitude of the dipole, and to the fourth power of the frequency. It can be used for such diverse problems as finding the rate of radiation from an excited atom, or from a radio antenna, if the latter can be approximated, as it often can, as an oscillating dipole.

5. Scattering of Light.—In Sec. 3 we have given an indication of the theory of the scattering of light. We can take up easily a special and important case of scattering. Let a plane wave fall on the type of dipole that we have assumed in Chaps. IV and IX as being responsible for the dispersion and dielectric properties of matter. We assume the dipole to be small in dimensions compared with the wave length, so that the electric field acting on the dipole may be taken to be $E_0 e^{j\omega t}$, disregarding the variations with position. Let the dipole have an equation of motion

$$
m \frac{d^2 x}{dt^2} + mg \frac{dx}{dt} + m\omega_0^2 x = eE_0 e^{j\omega t},
\tag{5.1}
$$

as in Eq. (1.2), Chap. IX. Then the dipole moment will be

$$Me^{j\omega t} = \frac{e^2}{m} \frac{E_0 e^{j\omega t}}{\omega_0^2 - \omega^2 + j\omega g}. \tag{5.2}$$

This is the oscillating dipole moment produced by the field.

The dipoles set into motion by the wave will emit light, which is scattered. The rate of emission by a single dipole is found by substituting the absolute value of the dipole moment in (5.2) into (4.2). Often the scattering is measured by the scattering cross section σ. This by definition is the area on which enough energy falls from the plane wave to equal the scattered intensity. Thus the rate of emission must equal σ times the Poynting vector of the plane wave, which is $\frac{1}{2} \sqrt{\epsilon_0/\mu_0}\, E_0^2$. Making the substitutions, we then find

$$\sigma = \frac{32\pi}{3} \frac{\omega^4}{[(\omega_0^2 - \omega^2)^2 + (\omega g)^2]} \left(\frac{e^2}{8\pi\epsilon_0 mc^2}\right)^2. \tag{5.3}$$

The quantity $(e^2/8\pi\epsilon_0 mc^2)$ is the quantity that we found, in Sec. 4, Chap. VIII, to be a classical radius R of the electron of charge e. Thus we find the interesting result that dimensionally the scattering cross section is determined by the square of this classical radius, or by the classical cross section of the electron, though with a numerical factor, and with a factor depending on frequency, which behaves quite differently in different parts of the spectrum. We shall now consider three important special cases of this scattering formula, holding for different frequencies:

a. The Rayleigh Scattering Formula.—This is what we have in the case where ω is small compared with ω_0. Since for ordinary atoms ω_0 is a frequency in the ultraviolet, we have this condition in the visible range of the spectrum. Then (5.3) becomes

$$\sigma = \frac{32\pi}{3} R^2 \left(\frac{\omega}{\omega_0}\right)^4. \tag{5.4}$$

The scattering is here proportional to ω^4, or to $1/\lambda^4$, where λ is the wave length. This is the Rayleigh scattering formula, developed to discuss the scattering of light by the sky. The proportionality to the inverse fourth power of the wave length means that the short blue and violet waves will be scattered by the air molecules much more than the long red ones, resulting in the blue color of the scattered light from the sky. The transmitted light thus has the blue removed and looks red, explaining the color near the sun at sunset.

b. The Thomson Scattering Formula.—In the other limiting case of X rays, when the frequency is large compared with ω_0, the scattering becomes

$$\sigma = \frac{32\pi}{3} R^2, \tag{5.5}$$

the Thomson scattering formula. This formula gives a scattering independent of the wave length, and is very important in discussing X-ray scattering by substances.

c. Resonant Scattering.—If ω is nearly equal to ω_0, it is evident that the denominator can become very small, resulting in very large scattering. This phenomenon can be much more conspicuous than the other two cases. Thus a bulb filled with sodium vapor, which has a natural frequency in the visible region, illuminated with light of this color, will scatter so much light that it appears luminous. This phenomenon is called "resonance scattering." Here, as with the absorption lines discussed in Chap. IX, we have to correlate our classical theory of electric dipoles with the quantum theory. Actual atoms scatter, as they absorb, at frequencies connected with quantum transitions between energy levels. However, the same description of the dipoles that is correct for describing absorption and dispersion is also applicable to scattering.

One observation regarding scattering is that, if the incident light is plane polarized, the dipoles will all vibrate along the direction of its electric vector. Thus there will be no intensity in the scattered light along this direction. The scattered light will have a maximum intensity at right angles, and it will be plane polarized. It was by experiments based on these facts that the polarization of X rays was first found.

6. Coherence and Incoherence of Light.—In the preceding paragraphs, we have calculated the scattering from a single dipole or a single atom, when a plane wave falls on it. It is found in practice that, in a gas containing N molecules per unit volume, the total intensity scattered by unit volume will be just N times that scattered by a single molecule. On the other hand, if the molecules are regularly arranged in a crystalline solid, there is practically no scattering The effect rather is the dispersion and absorption discussed in Chap IX. We must give closer consideration to the question of why this distinction exists, and why in particular the intensity of the scattered radiation from N molecules of a gas is N times that from a single molecule. Since the Maxwell equations are linear, the field vectors $\mathbf{E}$ and $\mathbf{H}$ satisfy the superposition principle, so that we should expect the total amplitude to be the sum of the amplitudes in the various waves, in which case the total intensity, being the square of the amplitude, would certainly not be the sum of the separate intensities. The key to this situation is found in the relations between the phases of the various waves that we are adding: if they all have fixed phases relative to each other, they are said to be "coherent," and the amplitudes add; whereas if they are in phases having random relations to each other they are "incoherent," and the intensities add.

To be more precise, let us consider the sum of a number of sinusoidal waves, all of the same frequency, but of different amplitude and phase:

$$\sum_k A_k \cos (\omega t - \alpha_k) = \left(\sum_k A_k \cos \alpha_k\right) \cos \omega t + \left(\sum_k A_k \sin \alpha_k\right) \sin \omega t.$$

If all the phases should be the same, say $\alpha_k = 0$, then the amplitudes of the cosine and sine terms will be $\sum_k A_k$ and 0, respectively, so that the amplitudes add, and the intensity is proportional to $\left(\sum_k A_k\right)^2$, or, if for instance there are N terms of equal amplitude, proportional to N^2 times the intensity of a single wave. On the other hand, the α's may be completely independent of each other, meaning that each α is equally likely to have any value between 0 and 2π, independent of the others. Then we can see that $\sum_k A_k \cos \alpha_k$ will be far less than $\sum_k A_k$, since we shall have just about as many terms with positive values of $\cos \alpha_k$ as with negative, and the terms will just about cancel.

The cancellation will not be complete, however, as we see if we compute the squares of the summations, which we must add to get the intensity. The square of the first summation, for instance, is

$$\left(\sum_k A_k \cos \alpha_k\right)^2 = \sum_k A_k^2 \cos^2 \alpha_k + \sum_{k \neq l} A_k A_l \cos \alpha_k \cos \alpha_l.$$

We must find the average of this, taking the α's as independent. That is, we must perform the operation of integrating each α from 0 to 2π, and dividing by 2π. When we do this, the terms $\cos^2 \alpha_k$ average to $\frac{1}{2}$, while the products of two independent α's average to zero, leaving $\frac{1}{2} \sum_k A_k^2$. The other summation gives an equal term, so that we find that the mean-square amplitude, or mean intensity, averaged over phases, is the sum of the individual intensities. This is the state of complete incoherence, in which for N waves the intensity is N times the intensity of a single wave, rather than N^2 as for the coherent case.   The cancellation of waves, then, while not complete, is more and more perfect as N increases, for N becomes a smaller and smaller fraction of N^2 as N increases.

We can now apply the idea of coherence to the scattering of light from a gas.   The phase of the wave at a point P, scattered by an atom at a, in Fig. 27, depends on the total path the light has traveled from the source to a, and from a to P.   Since the molecules of a gas have no fixed positions with respect to each other, these paths are in a random relation to each other, the phases are incoherent, and we are justified in adding intensities.   Such a procedure would not be allowed for example in discussing the scattering of X rays by crystals,

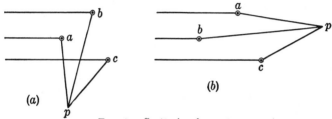

Fig. 27.—Scattering from atoms.

where the various atoms are in fixed lattice positions.   Indeed, here we do get interference, and it is just by studying the interference patterns so obtained that we obtain our information about the lattice structure of crystals.   Neither would the procedure be allowed in discussing the scattering from a gas in the same direction as the incident radiation, as in (b), Fig. 27.   For then the paths of the beams scattered from the various atoms are approximately equal, the waves are in phase, and they produce a resultant field at P proportional to the amplitude, rather than the intensity, of the incident wave. This scattered field can be shown to interfere with the incident wave in such a way that the resultant produces the refracted wave.   The close relation of our scattering formulas to the formulas for the index of refraction, therefore, becomes clear, and it is evident that our two problems of refraction and scattering, though we have treated them separately, are really parts of the same subject.   Scattering straight ahead produces refraction, and does not depend on exact placing of the molecules.   Scattering to the sides, on the other hand, does not occur unless the molecules have a random arrangement; then the intensity, not the amplitude, is proportional to the number of molecules.

We have just been considering the coherence of scattered light.

There is another aspect of coherence, related to this, which can be discussed in a similar way. This is the incoherence between different independent sources of light, even when they have the same frequency. It is based on the relation between coherence and the spectrum. The amplitude of a wave of light, as a function of time, is never exactly sinusoidal, but is really a much more complicated function. It is often desirable, however, to resolve such a function into a spectrum; that is, write it as a sum of sinusoidal waves of different frequency. This can be conveniently done by Fourier series, as is discussed in Appendix III.

To do this, we take a Fourier series with an extremely long period T, so long that all the phenomena we are interested in take place in a time short compared with T, so that we are not bothered by the periodicity of the series. Then, if our function is $f(t)$, we have

$$f(t) = \sum_n (A_n \cos \omega_n t + B_n \sin \omega_n t),$$

where

$$A_n = \frac{2}{T} \int_{-T/2}^{T/2} f(t) \cos \omega_n t \, dt, \qquad B_n = \frac{2}{T} \int_{-T/2}^{T/2} f(t) \sin \omega_n t \, dt,$$
$$\omega_n = \frac{2\pi n}{T}.$$

This gives an analysis into an infinite number of sine waves, with frequencies spaced very close together (on account of the very small size of $2\pi/T$). No actual, physical wave is then perfectly sinusoidal, in the sense of having but one term in this expansion with an amplitude different from zero. We shall show in a problem that even a perfectly sinusoidal wave that persists for only a finite length of time will have appreciable amplitudes for all those frequencies within a range $\Delta\omega$, equal in order of magnitude to the reciprocal of the time during which the wave persists, so that a sine wave of long lifetime will correspond to a sharp line in the spectrum, while a rapidly interrupted wave will give a broad line. This is observed experimentally in the fact that increasing the pressure of a gas, thereby making collisions more frequent and interrupting the radiating of the atoms, broadens the spectral lines.

The intensity is proportional to $f^2(t)$, or to the square of the summation over frequencies. Just as before, this square consists of terms like $A_n^2 \cos^2 \omega_n t$, and cross terms like $A_n A_m \cos \omega_n t \cos \omega_m t$. Instantaneously none of these terms are necessarily zero. But if we average over time, the terms of the first sort average to $A_n^2/2$, while those of the second sort average to zero. The final result, then, is that the time average intensity is the sum of the intensities of the various frequencies: $\overline{f^2(t)} = \frac{1}{2} \sum_n (A_n^2 + B_n^2)$. We are justified in considering the terms connected with a given n to be the intensity of light of that particular frequency in the spectrum, so that we have the theoretical method of determining the spectral analysis of any disturbance. And we see that the following statement is true: on a time average, sinusoidal waves of different frequencies are always incoherent, and never interfere.

It is known experimentally that light from two different sources never interferes; to get interference we must take light from a single source, split it into two beams, and allow these beams to recombine. If we regarded the sources as being monochromatic, it would be hard to see why this should be, for the amplitudes of two waves of the same frequency should add, rather than the intensities, and this is the essence of interference. But when we observe that each source really is represented by a Fourier series, the situation becomes plain. For two sources are always so different that their Fourier series will be entirely different. If we analyze both of them, the phase of the radiation of frequency ω_n from one will be entirely independent of the phase of the corresponding frequency from the other. Thus if we add, square, and average over this random relation between the phases of the two sources, the cross terms will cancel, and the intensities add. The randomness comes in this case, not in adding a great many terms of the same frequency, but in combining the terms of different frequencies, which are related in entirely independent ways in the two sources.

Problems

1. Discuss the weakening of sunlight because of scattering, as the light passes through the atmosphere. Assume that the molecules of the atmosphere have a natural frequency at 1,800 A (where absorption is observed). Let each molecule contain an electron of this frequency. Assume that the number of molecules is such as to give the normal barometric pressure. Find the fractional weakening of a beam due to scattering in passing through a sheet of thickness ds, and from this set up the differential equation for intensity as a function of the distance. Solve for the ratio of intensity to the intensity before striking the atmosphere, for the sun shining straight down, and for it shining at an angle of incidence of 60°. Constants: $e = 1.60 \times 10^{-19}$ coulomb, $m = 9.1 \times 10^{-31}$ kg, number of molecules in 1 gm-mol = 6.03×10^{23}.

2. A vibrating dipole radiates energy, and therefore its own energy decreases. Noting that the rate of radiation is proportional to the energy, set up the differential equation for the energy of the dipole as a function of the time. Find how long it takes the dipole to lose half its energy. Work out numerical values for the sort of dipole considered in Prob. 1.

3. Using the results of Prob. 2, find the equivalent damping term that would make the dipole lose energy at the same rate as the radiation. This damping is called the "radiation resistance."

4. Suppose we have an alternating current of maximum value I, in a vertical antenna of length L. Treating this as a dipole, find the total radiation. Show that the equivalent resistance necessary to produce the same power loss (the radiation resistance) is

$$R = \frac{2\pi}{3} \sqrt{\frac{\mu_0}{\epsilon_0}} \frac{L^2}{\lambda^2} \qquad \text{ohms.}$$

5. Show that the field of an oscillating dipole of moment $Me^{j\omega t}$, pointing along the axis of the spherical coordinates, can be derived from the following scalar and vector potentials:

$$\psi = -\frac{M}{4\pi\epsilon_0} \frac{d}{dr}\left[\frac{e^{j\omega(t-r/c)}}{r} \right] \cos\theta,$$

$$A_r = \frac{j\omega\mu_0 M e^{j\omega(t-r/c)}}{4\pi r} \cos\theta$$

$$A_\theta = -\frac{j\omega\mu_0 M e^{j\omega(t-r/c)}}{4\pi r} \sin\theta.$$

$$A_\varphi = 0.$$

6. The so-called "Hertz vector" arising from a dipole of moment $Me^{j\omega t}$, at a distance r from the dipole, is equal to

$$\mathbf{\Pi} = \frac{Me^{j\omega(t-r/c)}}{4\pi\epsilon_0 r},$$

and points in the direction of the dipole moment. Show that in terms of the Hertz vector the potentials can be found from the expressions

$$\mathbf{A} = \epsilon_0\mu_0 \frac{\partial\mathbf{\Pi}}{\partial t}, \qquad \psi = -\operatorname{div}\mathbf{\Pi}.$$

7. Show from the Hertz vector that the field at large distances from a dipole is given by

$$\mathbf{E} = \sqrt{\frac{\mu_0}{\epsilon_0}} \frac{1}{4\pi cr^3} \left\{ \mathbf{r} \times \left[\mathbf{r} \times \frac{d^2}{dt^2} \mathbf{M} e^{j\omega(t-r/c)} \right] \right\},$$

$$\mathbf{H} = -\frac{1}{4\pi cr^2} \left[\mathbf{r} \times \frac{d^2}{dt^2} \mathbf{M} e^{j\omega(t-r/c)} \right].$$

8. Find the spectrum of a disturbance that is zero up to $t = 0$, is sinusoidal until $t = T_0$, then is zero permanently. (*Hint:* Make the period T of the Fourier series indefinitely large compared with T_0.)

9. Find the spectrum of a disturbance that starts at $t = 0$, and is a sinusoidal damped wave after that. Show that the curve for intensity as a function of frequency has the same form as a resonance curve, in general, and that its breadth is connected with the damping constant in the same way. This illustrates an important principle: the emission and absorption spectra of the same substance are essentially equivalent. The resonance curve represents the absorption curve, because of the relation of forced oscillators and dispersion, whereas the damped wave is the emission. (*Hint:* Make the period T indefinitely large compared with the time taken for the oscillation to $1/e$th of its value.)

CHAPTER XIII

HUYGENS' PRINCIPLE AND GREEN'S THEOREM

Huygens' principle provides a well-known elementary method for treating the propagation of waves, and in this chapter we shall consider its mathematical background, showing its close connection with Green's theorem. The method is this: From each point of a given wave front, at $t = 0$, we assume that spherical wavelets originate. At time t, each wavelet will have a radius ct, and the envelope of these wavelets will form a new surface, which according to Huygens is simply the resulting wave front at this later time t. Thus, if the original wave front was a plane, it is easy to see that the final one will be a plane distant by the amount ct, while if it is a sphere, the final wave front will be a concentric sphere whose radius is larger by ct. In either case this construction gives us the correct answer, agreeing with the more usual methods of computation. The one difficulty is that our construction would give a wave traveling backward, as well as one traveling forward; the solution of this difficulty appears when we use the methods of the present chapter.

We may look at our process in a slightly different way, not used by Huygens, but developed early in the nineteenth century when the interference of light was being worked out by Young and Fresnel. Suppose that, instead of taking the envelope of all the spherical wavelets, we consider that each of these wavelets has a certain amplitude, consisting of a sinusoidal vibration. We then add these vibrations, just as if the wavelets were being sent out by interfering sources of light, and the resulting amplitude is taken to be that in the actual wave. This process can be shown to lead to essentially the same result, and it is this which can be justified theoretically. As a further generalization, it is not necessary to take the original surface to be a wave front; it can be any closed surface, so long as we allow the scattered wavelets to have the suitable phase and amplitude.

Our final result, then, is this: The disturbance at a point P of a wave field may be obtained by taking an arbitrary closed surface, and performing an integration over this surface. The contribution of a small element of area da of this surface equals the amplitude at P of a spherical wave starting from da at such a time that it reaches P at time t. The simplest form of spherical wave is one we did not mention in Chap. XII. It is a scalar solution of the wave equation, independent of angles, and depending only on r. The wave equation (1.3) of Chap. XII, in a case where u depends only on r, may be written

$$\frac{\partial^2(ru)}{\partial r^2} - \frac{1}{c^2}\frac{\partial^2(ru)}{\partial t^2} = 0,$$

which has solutions $u = \dfrac{f(t \pm r/c)}{r}$, where f is an arbitrary function. If we have a simple spherical wave of this type, we shall choose the $-$ sign, so as to represent a wave traveling out along the radius with velocity c, and of amplitude varying inversely as the distance. Now the contribution of an element da must surely be proportional to the disturbance at da, which we may call f (a function of time and position), and to da. Hence we have something like $\displaystyle\int \frac{f(t - r/c)}{r}\,da$ for the final result. We are thus led to a formula of this sort:

$$f \text{ (at a point } P) = \text{constant} \times \int \frac{f(t - r/c)}{r}\,da,$$

where the surface integral is over a surface surrounding P. This suggests the solution of Laplace's equation by Green's method, where we obtained the value of a function φ at an interior point of a region where $\nabla^2\varphi$ was zero as a surface integral over the boundary. As a matter of fact, an analogue to Green's theorem is the correct statement of Huygens' principle, and replaces the formula that we derived intuitively above, which is not exactly correct. We shall now proceed to set up this theorem, first introducing the retarded potentials, which are closely related to the problem.

1. The Retarded Potentials.—In Chap. VII, we introduced scalar and vector potentials, φ and $\mathbf{A}$, giving the electric and magnetic fields by the relations

$$\mathbf{E} = -\operatorname{grad}\varphi - \frac{\partial\mathbf{A}}{\partial t}$$

$$\mathbf{B} = \operatorname{curl}\mathbf{A}.$$

For these potentials, in empty space, we found the equations

$$\nabla^2\varphi - \frac{1}{c^2}\frac{\partial^2\varphi}{\partial t^2} = -\frac{\rho}{\epsilon_0}$$

$$\nabla^2\mathbf{A} - \frac{1}{c^2}\frac{\partial^2\mathbf{A}}{\partial t^2} = -\mu_0\mathbf{J}, \tag{1.1}$$

or d'Alembert's equation. We first ask how to get a solution of d'Alembert's equation analogous to the simple solution

$$\varphi = \frac{1}{4\pi\epsilon_0} \int \frac{\rho}{r} \, dv = -\frac{1}{4\pi} \int \frac{\nabla^2\varphi}{r} \, dv \qquad (1.2)$$

of Poisson's equation. We shall not carry through the proof of the solution, for that is rather complicated. But the essence of the solution of Poisson's equation is that we divide up all space into volume elements dv, and that $\rho \, dv/4\pi\epsilon_0 r$ is the potential of the point charge $\rho \, dv$ at a distance r. This potential, of course, is a solution of Laplace's equation, as is $1/r$, at all points except for $r = 0$, where the charge is located.

In a similar way, to solve d'Alembert's equation, we divide up our charge into small elements, and write the potential as the sum of the separate potentials of these small charges. The separate potentials must now be, except at $r = 0$, solutions of the wave equation. This means that, since any change of the charge will be propagated outward with the velocity c, the potential at a given point of space resulting from a particular charge cannot be derived from the instantaneous value of the charge, but must be determined, instead, by what the charge was doing at a previous instant, earlier by the time r/c required for the light to travel out from the charge to the point we are interested in. In other words, if $\rho(x,y,z,t)$ is the charge density at x, y, z at the time t, and r is the distance from x, y, z to x', y', z' where we are finding the field, we shall expect the potential of the charge in dv to be

$$\frac{\rho(x,y,z,t - r/c) \, dv}{4\pi\epsilon_0 r}$$

and for the whole potential we shall have

$$\varphi = \frac{1}{4\pi\epsilon_0} \int \frac{\rho(x,y,z,t - r/c) \, dv}{r}$$
$$= -\frac{1}{4\pi} \int \frac{[\nabla^2\varphi - (1/c^2)(\partial^2\varphi/\partial t^2)]_{t-r/c}}{r} \, dv. \qquad (1.3)$$

This solution is, as a matter of fact, correct. We have already seen that $\dfrac{f(t - r/c)}{r}$ is a solution of the wave equation, where f is any function, so that the integrand actually satisfies the wave equation, as in the earlier case $1/r$ satisfied Laplace's equation. The potential φ determined by this equation is called a "retarded potential," since any change in the charge is not instantaneously observable in the potential at a distant point, but its effect is retarded because of the finite velocity of light. The solution for each component of the vector potential is determined in an analogous manner.

2. Mathematical Formulation of Huygens' Principle.—In discussing the application of Green's theorem to the solution of Poisson's equation in a finite region of space, we proved in Sec. 6, Chap. III, that

$$\varphi = -\frac{1}{4\pi} \int \frac{\nabla^2 \varphi}{r} \, dv - \frac{1}{4\pi} \int \left(\varphi \operatorname{grad} \frac{1}{r} \cdot \mathbf{n} - \frac{1}{r} \operatorname{grad} \varphi \cdot \mathbf{n} \right) da, \quad (2.1)$$

the result of the last paragraph being the special case where the region of integration is infinite and the surface integral drops out. We now wish to find an analogous theorem for use with d'Alembert's equation. Here again we shall not give a real derivation, for this is complicated, but shall merely describe the formula that results, and show that it is plausible. We have already discussed the volume integral. In the surface integral, the first term gave the potential of a double layer of strength proportional to $\varphi/4\pi$, the second the potential of a surface charge of magnitude proportional to

$$\left(\frac{1}{4\pi}\right) \operatorname{grad} \varphi \cdot \mathbf{n}, \quad \text{or} \quad \left(\frac{1}{4\pi}\right)\left(\frac{\partial \varphi}{\partial n}\right).$$

Each of the terms, $\varphi \operatorname{grad} (1/r) \cdot \mathbf{n}$, or $\varphi[\partial(1/r)/\partial n]$ and $(1/r)(\partial \varphi/\partial n)$ is a solution of Laplace's equation since it represents the potential of certain charges.

In our case of the wave equation, the formula has two corresponding terms: one giving the potential of a double layer, the other of a surface charge. But now the charges change with time, so that we must use solutions of the wave equation in the integral. We have already seen that the solution of the wave equation corresponding to $1/r$ is $f(t - r/c)/r$; hence we expect the second term to be replaced by $-(1/r)(\partial \varphi/\partial n)_{t-r/c}$, where this means that the partial derivative, which is now a function of time as well as of position on the surface, is to be computed, not at t, but at $t - (r/c)$. Similarly corresponding to $\partial(1/r)/\partial n$, the difference of the potentials of two equal and opposite point charges at neighboring points of space, we have

$$\frac{\partial}{\partial n} \left[\frac{f(t - r/c)}{r} \right].$$

Remembering that in differentiating with respect to n we must regard r as a variable each time it occurs, this is

$$f\left(t - \frac{r}{c}\right)\frac{\partial(1/r)}{\partial n} + \frac{1}{r}\frac{\partial}{\partial n}\left[f\left(t - \frac{r}{c}\right)\right]$$

$$= -\frac{\cos(n,r)}{r}\left[\frac{f(t - r/c)}{r} + \frac{1}{c}\frac{\partial f(t - r/c)}{\partial t}\right]$$

where in the last term we have used the relation

$$\frac{\partial f(t - r/c)}{\partial n} = \frac{df(t - r/c)}{d(t - r/c)}\frac{\partial(t - r/c)}{\partial n}$$

$$= \frac{\partial f(t - r/c)}{\partial t}\left(-\frac{1}{c}\frac{\partial r}{\partial n}\right)$$

$$= -\frac{1}{c}\cos(n,r)\frac{\partial f(t - r/c)}{\partial t}.$$

We should therefore expect to have

$$\varphi = -\frac{1}{4\pi}\int\frac{[\nabla^2\varphi - (1/c^2)(\partial^2\varphi/\partial t^2)]_{t-r/c}}{r}\,dv$$

$$+ \frac{1}{4\pi}\int\frac{1}{r}\left\{\left[\frac{1}{c}\left(\frac{\partial\varphi}{\partial t}\right)_{t-r/c} + \frac{\varphi(t - r/c)}{r}\right]\cos(n,r)\right.$$

$$\left. + \left(\frac{\partial\varphi}{\partial n}\right)_{t-r/c}\right\}\,da. \quad (2.2)$$

This, as a matter of fact, is the correct formula. The first term represents the potential due to all the charge within the volume; if there are no sources of radiation within this volume, the volume integral is then zero, and that is the usual case with optical applications. The surface integral represents the remaining potential as arising from a distribution of charge and double distribution about the surface, each surface element sending out a wavelet that on closer examination proves to be the Huygens' wavelet we are interested in. Thus, starting from Green's theorem and d'Alembert's equation, we have arrived at a mathematical formulation of Huygens' theorem.

To give a suggestion of the rigorous proof of this formula, we could proceed as follows: First, we notice that φ defined by this integral satisfies the wave equation; for since each term of the integrand separately is a solution, the sum must also be. Now it follows from this, although we have not proved it, that if the solution reduces to the correct boundary values at all points of the boundary, the solution must be the correct one, the reason being essentially that the boundary values determine a solution uniquely, so that, if we have one solution of the equation with the right boundary values, it must be the only correct solution. We must then show that the φ defined by the integral actually has the correct boundary values. This could be done by a more careful treatment, and we should then have a demonstration of the formula. The more conventional proof, however, is a fairly direct though complicated application of Green's theorem.

We shall now take our general formula (2.2), and apply it to the cases we meet in optics, showing that it reduces to something like the formula that we derived earlier intuitively. We suppose that light is emitted by a point source, and that the value of some quantity connected with it, and satisfying the wave equation (one of the components of the fields or potentials—they all satisfy the same relations) has the form $\dfrac{A\,e^{j\omega(t-r_1/c)}}{r_1}$, where r_1 is the distance from the source to the point where we wish to find the disturbance. Then we wish to get the disturbance at P, not by direct calculation, but by using Huygens' principle. Suppose we take a closed surface. This surface can surround either the source, or the point P where we wish the disturbance. In any case, we define **n** as the normal pointing out of the part of the space in which P is located. At a point of the surface, $\varphi = \dfrac{A\,e^{j\omega(t-r_1/c)}}{r_1}$, where r_1 is the distance from the source to the point on the surface.

We then have, if r is the distance from P to a point on the surface,

$$\varphi\left(t - \frac{r}{c}\right) = \frac{A\,e^{j\omega[t-(r+r_1)/c]}}{r_1}$$

$$\left(\frac{\partial \varphi}{\partial t}\right)_{t-r/c} = \frac{j\omega A\,e^{j\omega[t-(r+r_1)/c]}}{r_1}$$

$$\left(\frac{\partial \varphi}{\partial n}\right)_{t-r/c} = -A\,\cos\,(n,r_1)\left(\frac{1}{r_1} + \frac{j\omega}{c}\right)\frac{e^{j\omega[t-(r+r_1)/c]}}{r_1}.$$

Thus finally, substituting in (2.2) and rearranging terms,

$$\varphi = \frac{1}{4\pi}\int \frac{A}{rr_1}\,e^{j\omega[t-(r+r_1)/c]}\left[\left(\frac{1}{r} + \frac{j\omega}{c}\right)\cos\,(n,r)\right.$$
$$\left. - \left(\frac{1}{r_1} + \frac{j\omega}{c}\right)\cos\,(n,r_1)\right]\,da. \quad (2.3)$$

In this formula, as in Chap. XII, we have two sorts of terms, some significant at small, others at large values of r and r_1. We easily see that, if r and r_1 are large compared with a wave length, as is always the case in optics, the only terms we need retain are those in $j\omega/c$. Hence to this approximation

$$\varphi = \int \frac{jA}{2\lambda r_1 r}\,e^{j\omega[t-(r+r_1)/c]}[\cos\,(n,r) - \cos\,(n,r_1)]\,da. \quad (2.4)$$

This final form suggests our earlier, intuitive formulation of Huygens' principle.   The incident amplitude at da is $\dfrac{A\,e^{j\omega(t-r_1/c)}}{r_1}$.   Now we set up, starting from da, a wavelet whose amplitude is this value, retarded by the amount r/c, divided by r, and multiplied by the factor $(j/2\lambda)[\cos\,(n,r) - \cos\,(n,r_1)]\,da$.   This is just what we should expect, except for the last factor.   The term j introduces a change of phase of 90°, not present in Huygens' form of the principle, but necessary. The term $\cos\,(n,r) - \cos\,(n,r_1)$ makes the wavelets have an amplitude that depends on angle.   When r and r_1 are in opposite directions, which is the case when the surface is between the source and P, the factor approaches 2; whereas, when r and r_1 are parallel, and the surface is beyond P, it becomes zero.   This means that the wavelets do not travel backward, thus removing the difficulty noticed earlier in Huygens' method.   The wavelets have an amplitude depending on their wave length, decreasing for the longer wave lengths.

3. Integration for a Spherical Surface by Fresnel's Zones.—Let us now carry out our integration, and verify Huygens' method, in a simple case.   We take the surface to be a sphere, surrounding the source, and therefore a wave front.   We note that n is the inner normal of the sphere.   Thus r_1 is constant all over the sphere, and $\cos\,(n,r_1) = -1$ at all points, so that the formula simplifies to

$$\varphi = \frac{jA\,e^{j\omega(t-r_1/c)}}{2\lambda r_1} \int \frac{e^{-jkr}}{r}\,[\cos\,(n,r) + 1]\,da,$$

where $k = \omega/c = 2\pi/\lambda$.   Now suppose we introduce, as a coordinate on the sphere, the distance r from the point P; that is, we cut the sphere with spheres concentric with P, laying off zones between them, as in Fig. 28.   We can easily get the area between r and $r + dr$, and hence

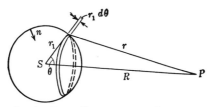

Fig. 28.—Construction for Fresnel's zones on a sphere surrounding the source.

the element of area.   Take as an axis the line joining the source and the point P, and consider a zone making an angle between θ and $\theta + d\theta$ with the axis.   The area of the zone is $2\pi r_1^2 \sin\theta\,d\theta$.   But now by the law of cosines, if R is the distance from the source to P,

$$r^2 = R^2 + r_1^2 - 2Rr_1 \cos \theta,$$

and differentiating, $2r \, dr = 2Rr_1 \sin \theta \, d\theta$. Hence for the area of the zone we have $\dfrac{2\pi r r_1}{R} dr$. Introducing this, we have

$$\varphi = \frac{j\pi A \, e^{j\omega(t-r_1/c)}}{\lambda R} \int_{r_{\min}}^{r_{\max}} e^{-jkr}[\cos \, (n,r) + 1] \, dr,$$

where $r_{\min} = R - r_1$, $r_{\max} = R + r_1$.

To carry out this integration, we use a device called "Fresnel's zones," giving us an approximate value in a very elementary way. Beginning with $r_{\min}$, we take a set of zones such that the outer edge of each corresponds to a value of r just half a wave length greater than the inner edge. The contributions of successive zones will almost exactly cancel. The integral, then, consists of a sum of terms, say $s_1 - s_2 + \cdots + s_n$, where the magnitudes of $s_1, s_2, \cdots$, vary only slightly from one to the next. Now it is true in general that in such a series the sum is approximately half the sum of the first and last terms. We can see this as follows: We group the terms

$$\frac{s_1}{2} + \left(\frac{s_1}{2} - s_2 + \frac{s_3}{2}\right) + \cdots + \left(\frac{s_{n-2}}{2} - s_{n-1} + \frac{s_n}{2}\right) + \frac{s_n}{2}.$$

Now, because of the slow variation of magnitude, we have very nearly $s_k = \dfrac{s_{k-1} + s_{k+1}}{2}$. If this were so, however, each of the parentheses would vanish, leaving only $\dfrac{s_1 + s_n}{2}$.

In our case, the contribution of the first zone is to be considered, but that of the last zone is practically zero, because of the factor $\cos \, (n,r) + 1$, so that the result is half the first zone. Now, in the first zone, $\cos \, (n,r) + 1$ is so nearly equal to 2 that we can take it outside the integral, obtaining

$$\begin{aligned}
\varphi &= \frac{\pi j}{\lambda} \frac{A \, e^{j\omega(t-r_1/c)}}{R} \int_{R-r_1}^{R-r_1+\lambda/2} e^{-jkr} \, dr \\
&= -\frac{A \, e^{j\omega(t-r_1/c)}}{2R} \left(e^{-jkr}\right) \Big|_{R-r_1}^{R-r_1+\lambda/2} \\
&= \frac{A \, e^{j\omega(t-r_1/c)}}{R} e^{-jk(R-r_1)} \\
&= \frac{A \, e^{j\omega(t-R/c)}}{R},
\end{aligned} \tag{3.1}$$

the correct value.

In the derivations of this chapter we have traveled in a very roundabout way to reach a very obvious result. We naturally ask, what is Huygens' principle good for, aside from a mathematical exercise? The answer is found in problems of diffraction. There one has certain opaque screens, with holes in them, and a light wave falling on them. If the light comes from a point source, geometrical optics would tell us that the shadow of the screen would have perfectly sharp edges. But actually this is not true; there are light and dark fringes around the edge of the shadow. If the shadow is observed at a greater and greater distance, these fringes get proportionally larger and larger, until they entirely fill the image of the hole. Finally at great distances the fringes grow in size until the resulting pattern has no resemblance at all to the geometrical image. There are then two general sorts of diffraction: first, that in which the pattern is like the geometrical image, but with diffuse edges, and which is called "Fresnel diffraction"; secondly, that in which the pattern is so extended that it has no resemblance to the geometrical image, and which is called "Fraunhofer diffraction." Both types of diffraction, as well as the intermediate cases, can be treated by using Huygens' principle.

4. Huygens' Principle for Diffraction Problems.—Suppose that light from a point source falls on a screen containing apertures, and that we wish the amplitude at points behind the screen. Then we surround the point P, where we wish the field, by a surface consisting of the screen, and of a large surface, perhaps hemispherical, extending out beyond P, and enclosing a volume completely. We apply Huygens' principle to the surface. In doing so we assume (1) that the amplitude of the incident wave, at points on the apertures, is the same that it would be if the screen were absent; and (2) that immediately behind the screen, and at points of the hemispherical surface as well, the amplitude is zero, the wave being entirely cut off by the screen. This is, of course, an approximation, since at the edge of a slit, for example, the amplitude of the wave does not suddenly jump from zero to a finite value. The exact treatment is exceedingly difficult, but in the few cases for which it has been worked out, it substantiates our approximations, if the dimensions of the aperture are large compared with the wave length, showing that they lead to a correct qualitative understanding of the phenomena, though there are quantitative deviations.

To find the disturbance at P, then, we integrate over the surface, but set the integrand equal to zero, except at the openings of the screen, obtaining

$$\varphi = \int \frac{jA}{2\lambda} \frac{1}{rr_1} e^{j\omega[t-(r+r_1)/c]}[\cos(n,r) - \cos(n,r_1)]\,da,$$

the integral being over the openings. We note that only the edges of the openings are significant, the shape of the screen away from the openings being unimportant. Now let us assume, as is almost always true in practice, that the distances r_1 and r, from source to screen and from the screen to P, are large compared with the dimensions of the holes. Then $1/rr_1$ and $[\cos(n,r) - \cos(n,r_1)]$ are so nearly constant over the aperture that we may take them outside the integral, replacing r and r_1 by mean values $\bar{r}$ and $\bar{r}_1$. If in addition we write $r + r_1$ in the exponential as $\bar{r} + \bar{r}_1 + r' + r_1'$, where r' and r_1' are the small differences between r and r_1, and their values at some mean point of the aperture, we have finally

$$\varphi = \frac{jA}{2} \frac{1}{\bar{r}\bar{r}_1}[\cos(n,\bar{r}) - \cos(n,\bar{r}_1)]e^{j\omega[t-(\bar{r}+\bar{r}_1)/c]} \int e^{-jk(r'+r_1')}\,da. \quad (4.1)$$

The whole factor outside the integral may be taken as a constant factor so that, if we are interested only in relative intensities, we may leave it out of account. We finally have a sinusoidal vibration of which the amplitudes of the components of the two phases are proportional to $C' = \int \cos k(r' + r_1')\,da$, and $S' = \int \sin k(r' + r_1')\,da$. Hence the intensity is proportional to $C'^2 + S'^2$, and our task is to compute this value.

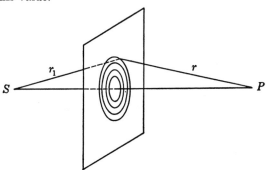

Fig. 29.—Fresnel's zones on a plane.

By using Fresnel's zones, one can see qualitatively the explanation of the diffraction fringes, particularly in Fresnel diffraction. Suppose that we join the source S and a point P with a straight line, as in Fig. 29, and consider the point of the screen cut by this line, a point for which $r + r_1$ has a minimum value. Let us surround this point by successive closed curves in which $r + r_1$ differs from its minimum value by successive whole numbers of half wave lengths. It is not hard to see that these curves will be the intersections with the screen of a set of ellipsoids of revolution, whose foci are S and P. Hence if the line SP is approximately normal to the screen, the curves will be approximately circles. Successive zones included between successive curves will propagate light differing by a half wave length from their neighbors. Now on the screen we may imagine the pattern of zones, and also the apertures. The whole nature of the diffraction depends on what zones are uncovered, and can transmit light, and what ones are obscured by the screen. We may distinguish three cases, shown in Fig. 30:

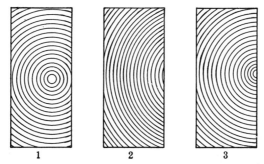

Fig. 30.—Fresnel's zones and rectangular aperture. (1) Directly in path of light. (2) In geometrical shadow. (3) On edge of shadow.

1. The center of the system of zones lies well inside the aperture. The central zone is entirely uncovered, as are a number of the others. As we get to larger zones, we shall come to one of which a small part is covered; then one that is more covered; and so on, until finally we come to one only slightly uncovered; and then the rest are entirely obscured. Now we can write our integral, as in Sec. 3, as a sum of integrals over the successive zones. As before, these contributions will decrease very gradually from one zone to the next. When we reach the zones that are obscured, the decrease will become a little more rapid, but not so much as to interfere with the argument. We can still write the whole thing as half the sum of the first and the last zones. In our case, the last zone that contributes has a negligibly small area exposed, so that it contributes practically nothing, and the whole integral is half the first zone. But this gives just the intensity we should have in the absence of the screen.

2. The center of the zone system is well behind the screen (P is in the geometrical shadow). Then the first few zones are obscured. A certain zone begins to be uncovered, until finally some zones are uncovered to a considerable extent. Large zones become obscured again, however. Thus in our sum, although there are terms different from zero, both the first and the last terms are zero, so that the sum is zero. The intensity well inside the geometrical shadow is zero.

3. The center of the zone system is near the edge of the screen. Then the first zone may be partly obscured, so that there is some intensity, but not so great as without the screen. Or the first zone may be entirely uncovered, but the next ones partly obscured. In these cases, the contributions from the successive zones may differ so much that our rule of taking the first and last terms is no longer correct. It is possible for the whole amplitude to be more than half the first zone, so that the intensity is actually greater than without the screen. As we move into the geometrical image from the shadow, it turns out that there is a periodic fluctuation, because of the uncovering of successive zones, and this explains the diffraction fringes.

Problems

1. Try to carry out exactly the integration that we did approximately by using Fresnel's zones.

2. The source is at infinity, so that a wave front is a plane. Set up Fresnel's zones, and find the breadth of the nth zone, and its area.

3. A plane wave falls on a screen in which there is a circular hole. Investigate the amplitude of the diffracted wave at a point on the axis, showing that there is alternate light and darkness either as the radius of the hole increases, or as the point moves toward or away from the screen. (*Suggestion:* The integral consists of a finite number of zones.)

4. A plane wave falls on a circular obstacle. Show that at a point behind the obstacle, precisely on the axis, there is illumination of the same intensity that we should have if the obstacle were not there. Explain why this would not hold for other shapes of the obstacle.

5. Take a few simple alternating series, such as, $\frac{1}{2} - \frac{1}{3} + \frac{1}{4} - \frac{1}{5} + \cdots$, $\frac{1}{2} - \frac{1}{4} + \frac{1}{6} - \cdots$, $\frac{1}{2}^2 - \frac{1}{3}^2 + \frac{1}{4}^2 - \cdots$, and so forth, and find whether our theorem about the sum of a number of terms is verified for them. In doing this, it may be necessary to start fairly well out in the series, so as to satisfy our condition that successive terms differ only slightly in magnitude.

6. Prove the statement that the boundaries of Fresnel's zones are the intersection of the screen with ellipsoids of revolution whose foci are the source and the point P. What happens to these ellipsoids as the source is removed to infinity?

7. A "zone plate" is made by constructing 20 Fresnel zones on a plane glass plate for a source at infinity and blocking off the light from every other zone. This zone plate, when illuminated at normal incidence by a plane wave, acts as a lens and produces a bright spot on its axis at a distance of 1 m from the plate.

If the incident light is monochromatic and of wave length 5,000 A, compute the radius of the zone plate and the intensity at the bright spot relative to its value in the absence of the zone plate. At what other points on the axis will there be intensity maxima?

8. Consider a line source of light and a plane observation screen parallel to and at a finite distance from the line source. The wave fronts are cylindrical. Show that the Fresnel zones on a wave front at a distance d from the source are strips parallel to the line source, and compute the angles subtended by these zones at the source. Suppose a straight-edged obstacle is inserted between the source and the screen, cutting off all the light below the plane perpendicular to the screen and passing through the light source. Discuss the variation of light intensity as a function of position on the screen, and obtain approximate expressions for the location of the maxima and minima.

CHAPTER XIV

FRESNEL AND FRAUNHOFER DIFFRACTION

In the present chapter we proceed to the mathematical discussion of Fresnel and Fraunhofer diffraction, based on the methods of Huygens' principle derived in Chap. XIII. The problems that we take up are Fresnel and Fraunhofer diffraction through a slit; Fraunhofer diffraction through a circular aperture; and the diffraction grating, an example of Fraunhofer diffraction. In Eq. (4.1) of the preceding chapter, we saw that the essential step in computing the diffraction pattern is the evaluation of the integral

$$\int e^{-ik(r+r_1)} \, da,$$

where $k = 2\pi/\lambda$, and where the integration is over the aperture of the screen, da is an element of surface in the aperture, r is the distance from the source to the element da, and r_1 the distance from the element to the point P where the field is being computed. If the incident wave is a plane wave, and the plane of the aperture is a wave front, then r is the same for all elements, and the factor e^{-ikr} can be canceled out of the integral. The remaining integral, $\int e^{-ikr_1} \, da$, represents the sum at P of the amplitudes of spherical waves of equal intensity and phase starting from all points of the aperture. It is the interference of these waves which produces the diffraction pattern.

1. Comparison of Fresnel and Fraunhofer Diffraction.— The two types of diffraction, Fresnel and Fraunhofer, arise from observing the pattern near to, or far from, the screen. Let the normal to the screen be the z axis, as in Fig. 31, and let the screen containing the aperture be at $z = 0$. The light passing through the aperture is caught on a second screen at $z = R$. Physically, the diffraction pattern has the following nature: close to the aperture, the light passes along the z axis as a column or cylinder of illumination, of cross section identical with the aperture, so that, if the screen at R is close to the aperture, the illuminated region will have the same shape as the aperture, and we speak of rectilinear propagation of the light. As R increases, however, the column of light begins to acquire fluctuations of intensity near its boundaries, so that the pattern on the screen has fringes around the edges. This phenomenon is the Fresnel diffraction.

The size of the Fresnel fringes increases proportionally to the square root of the distance R. Thus Fig. 32 shows, in its upper diagram, the slit, parallel column of light, and parabolic lines starting from the edges of the slit, indicating the position of the outer bright fringe of the Fresnel pattern, if we are sufficiently near to the slit. As R becomes larger, the fringes become so large that there are only one or two in the pattern of the aperture, and the pattern shows but small resemblance to the shape of the aperture, though it still is of roughly

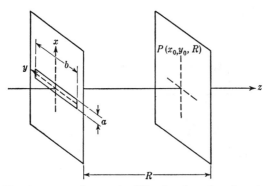

Fig. 31.—Aperture and screen for diffraction through rectangular slit.

the same dimensions. With further increase of R, we finally enter the region of Fraunhofer diffraction. Here the beam of light, instead of consisting of a luminous cylinder, resembles more a luminous cone indicated by the diverging dotted lines in the top diagram of Fig. 32. Thus the Fraunhofer pattern becomes larger and larger as R increases, being in fact proportional to R, so that we can describe it by giving the angles rather than distances between different fringes.

Often Fraunhofer diffraction is observed, not by placing the screen at a great distance, but by passing the light through a telescope focused on infinity. Such a telescope brings the light in a given direction to a focus at a given point of the field. Thus it separates the different Fraunhofer fringes, since each of these goes out from the source in a particular direction. In Fig. 32, diffraction patterns are shown indicating the transition from Fresnel to Fraunhofer diffraction. Pattern a illustrates the Fresnel pattern for one edge of an infinitely wide slit. Patterns b to g represent the actual diffraction patterns from the slit, at distances indicated in the upper diagram.

These patterns are all drawn to the same scale. They are drawn from a slit five wave lengths wide, for the sake of getting the figure on a diagram of reasonable scale. If the wave length were shorter, then for the same slit the distances would be stretched out to the

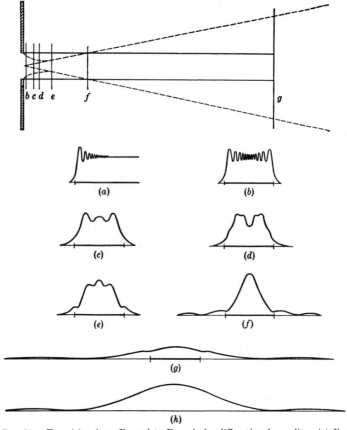

Fig. 32.—Transition from Fresnel to Fraunhofer diffraction for a slit. (*a*) Fresnel pattern for edge of infinitely wide slit. (*b*)–(*g*) Actual diffraction patterns from slit, at distances indicated in upper diagram. (*h*) Fraunhofer pattern.

right, and the Fraunhofer pattern would correspond to smaller angular deflections. This would be necessary to bring the Fresnel cases far enough from the slit so that our approximations would be really applicable. Finally, in *h*, we give the limiting Fraunhofer pattern, not drawn to scale.

Let coordinates in the plane of the aperture be x, y, and in the plane of the screen at R let the coordinates be x_0, y_0, as in Fig. 31. Then, if the element of area is at x, y, 0, and the point P at x_0, y_0, R, the distance r_1 between them is

$$r_1 = \sqrt{(x_0 - x)^2 + (y_0 - y)^2 + R^2}.$$

The integration cannot be performed with this expression for r_1, and Fresnel and Fraunhofer diffraction lead to two different approximate methods of rewriting r_1, leading to different methods of evaluating the integral. We can see the relation of these two methods most clearly from Fig. 33, in which r_1 is plotted as a function of $x_0 - x$, for the special case where $y_0 - y$ is zero. The resulting curve is a hyperbola.

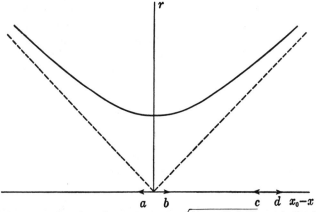

Fig. 33.—r_1 as function of $x_0 - x$. $r_1 = \sqrt{(x_0 - x)^2 + R^2}$. r_1 is the distance from a point of the aperture to a point on the screen; $x_0 - x$ is the difference between the x coordinates of the points.

Now in all ordinary cases, R is large compared with the dimensions of the aperture. That is, the range of abscissas representing the dimensions of the aperture (from $x_0 - x_1$ to $x_0 - x_2$, if x_1 and x_2 are the extreme coordinates of the aperture) is small compared with the distance R, the intercept of the hyperbola on the axis of ordinates. The two cases are now represented by the ranges ab and cd of abscissas, respectively. In the first, $x_0 - x_1$ and $x_0 - x_2$ are separately small, as well as their difference, and this means that the point P is almost straight behind the aperture, in the region where the Fresnel diffraction pattern occurs. In the second, x_0 is large, of the same order of magnitude as R, showing that we are examining the pattern at a considerable angle to the normal, as we do in the Fraunhofer case.

The two approximate methods can now be simply described from the curve: for Fresnel diffraction, we approximate the hyperbola near its minimum by a parabola; for Fraunhofer diffraction, we approximate it farther out by a straight line. In the first case, assuming R to be large compared with $(x_0 - x)$, we have by the binomial expansion

$$r_1 = R + \frac{1}{2} \frac{(x_0 - x)^2}{R} + \cdots ,$$

or, including the terms in y,

$$r_1 = R + \frac{1}{2} \frac{(x_0 - x)^2 + (y_0 - y)^2}{R} + \cdots . \tag{1.1}$$

In this case, in the notation of Eq. (4.1) of Chap. XIII, we take $\bar{r} = R$, so that r' is the remaining term of (1.1). For Fraunhofer diffraction, on the other hand, we have $x_0 \gg x$. Then we can write $r_1^2 = (x_0^2 + y_0^2 + R^2) - 2(xx_0 + yy_0) + x^2 + y^2$, and we can neglect the terms $x^2 + y^2$. If we let $R_0^2 = x_0^2 + y_0^2 + R^2$, where R_0 measures the distance from the center of the aperture to the point P, we can use a binomial expansion, obtaining

$$r_1 = R_0 - \frac{xx_0 + yy_0}{R_0} - \cdots . \tag{1.2}$$

In this case we take $\bar{r} = R_0$, so that r' is the remaining term of (1.2). Letting $x_0/R_0 = l$, $y_0/R_0 = m$, the direction cosines of the direction from the center of the aperture to P, we have

$$r' = -(lx + my) \cdots ,$$

involving the position on the screen only through the angles, so that we see at once that the pattern will travel outward radially from the aperture.

2. Fresnel Diffraction from a Slit.—Let the aperture be a slit, extending from $x = -(a/2)$ to $x = a/2$, and from $y = -(b/2)$ to $b/2$. We assume a to be small, b comparatively large, as in Fig. 31, so that it is a long narrow slit. Using the results of (1.1), our integral is

$$\int e^{-jkr'} \, da = \int e^{-\pi j[(x-x_0)^2 + (y-y_0)^2]/R\lambda} \, da.$$

This can be immediately factored into

$$\int_{-b/2}^{b/2} e^{-\pi j(y-y_0)^2/R\lambda} \, dy \int_{-a/2}^{a/2} e^{-\pi j(x-x_0)^2/R\lambda} \, dx.$$

Since these two integrals are of the same form, we can treat just one of them. This will prove to give fringes parallel to one set of axes. The whole pattern is then simply the combination of the two sets of fringes.   The single integral, for instance, the one in x, has a real part, and an imaginary part (with sign changed), equal to

$$\int_{-a/2}^{a/2} \cos \frac{\pi(x - x_0)^2}{R\lambda}\, dx \qquad \text{and} \qquad \int_{-a/2}^{a/2} \sin \frac{\pi(x - x_0)^2}{R\lambda}\, dx. \qquad (2.1)$$

It is customary in these integrals to make a change of variables: $\dfrac{(x - x_0)^2}{R\lambda} = \dfrac{u^2}{2}.$ Then the integrals become $\sqrt{R\lambda/2}$ times C and S, respectively, where $C = \displaystyle\int_{u_1}^{u_2} \cos \frac{\pi}{2} u^2\, du,\ S = \int_{u_1}^{u_2} \sin \frac{\pi}{2} u^2\, du$, and where $u_1 = \dfrac{x_0 - a/2}{\sqrt{R\lambda/2}},\ u_2 = \dfrac{x_0 + a/2}{\sqrt{R\lambda/2}}.$ These integrals are called "Fresnel's integrals." They cannot be explicitly evaluated, but their values have been computed by series methods.

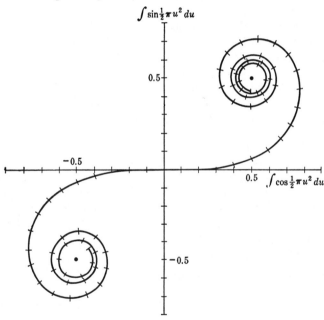

Fig. 34.—Cornu's spiral.   The points of the spiral marked by cross bars correspond to increments of 0.1 unit in u.

To discuss Fresnel's integrals, let us plot the indefinite integral $\displaystyle\int_0^u \cos \frac{\pi}{2} u^2\, du$ as abscissa, $\displaystyle\int_0^u \sin \frac{\pi}{2} u^2\, du$ as ordinate, of a graph, as in Fig. 34.   Then it is not hard to see that the resulting curve is a spiral, which is known as "Cornu's spiral." To see this, we can first compute the slope. This is the differential of the ordinate, over the differential of the abscissa, or

$$\frac{\sin{(\pi/2)u^2}}{\cos{(\pi/2)u^2}} = \tan{\frac{\pi}{2}}u^2.$$

Thus, when u^2 increases by 4, the tangent of the curve swings around a complete cycle, and comes back to its initial value. Each point of the spiral corresponds to a particular value of u. We can show at once that the difference of u between two points is simply the length of the curve between the points. We show this for an infinitesimal element of the curve. The square of the element of length, ds^2, is equal to the sum of the squares of the differentials of abscissa and ordinate, or is $\cos^2{[(\pi/2)u^2]}\,du^2 + \sin^2{[(\pi/2)u^2]}\,du^2$. Hence $ds = du$, and we can integrate to get $s = u_2 - u_1$. From this fact we can make sure of the spiral nature of the curve. For one turn corresponds to an increase of u^2 by 4. That is, if u' and u'' are the values at the two ends, $u''^2 = u'^2 + 4$. This is $u''^2 - u'^2 = 4$, $(u'' - u')(u'' + u') = 4$, $u'' - u' = 4/(u'' + u')$. The difference $u'' - u'$ is, however, simply the length of the turn, so that as we go farther along, the turns become smaller and smaller, so that they eventually become zero, which is characteristic of a spiral. It is plain that the spiral is symmetric in the origin, having two points, for $u = \pm \infty$, for which it winds up on itself.

Let us take our spiral, mark on it the positions u_1 and u_2 corresponding to the limits of our integral, and draw the straight line connecting these points. The length of this line will then be proportional to the amplitude of the disturbance, and its square to the intensity. This is easy to see: the horizontal component of the line is just C, and the vertical component S, so that the square of its length is $C^2 + S^2$. Knowing this, we can easily discuss the fluctuations of intensity, as seen in Fig. 32. As x_0 changes, it is plain that u_1 and u_2 increase together, their difference remaining fixed and equal to $\dfrac{a}{\sqrt{R\lambda/2}}$. Thus essentially we have an arc of this length, sliding along the spiral, and the intensity is measured by the square of the chord between the ends of this arc.

Now when x_0 is large and negative, the arc is wound up on itself, so that its ends practically meet, and the intensity is zero. This is the situation in the shadow. As x_0 approaches the value $-a/2$, however, u_2 approaches zero, so that one end of the arc has reached the center of the figure. There are two quite different cases, depending on whether $u_2 - u_1$ is large or small. If it is large (a large slit and relatively short distance R and small wave length), then u_1 will not be unwound much at this point. The chord will then be half the value between the two end points of the spiral, and the intensity will be one-fourth its value without the screen, and will have increased uniformly in coming out of the shadow. As we go farther along the x direction, however, the arc will begin to wind up on the other half of the spiral, producing alternations of intensity at the end of the shadow. Then for a while u_2 will be nearly at one end of the spiral, u_1 at the other, so that the intensity for some distance will be nearly constant, and the same that we should have without the slit. This is the illuminated region directly behind the slit. Finally we approach the other boundary, and u_1 commences to unwind. We then go through the same process in the opposite order.

The other quite different case comes when $u_2 - u_1$ is small, which is the case for small slit, or large wave length or distance. Then there is never a time when u_1 is on one branch of the spiral and u_2 on the other. All through the central part of the pattern, therefore, there are no fluctuations of intensity. Such fluctuations come only far to one side or the other. They come about in this way: At some places in the pattern, the arc is long enough to wind up for a whole number of turns, and the chord is practically zero, while at other places it winds up for a whole number plus a half, and the chord has a maximum. The resulting fringes are the Fraunhofer fringes which we shall now discuss by a different method.

3. Fraunhofer Diffraction from a Slit.—Using the approximation (1.2), our integral for Fraunhofer diffraction is $e^{-jkR_0} \int e^{2\pi j(lx+my)/\lambda} \, da$. The first term, as in Fresnel diffraction, contributes nothing to the relative intensities, and may be discarded. We then have

$$\int e^{2\pi j(lx+my)/\lambda} \, da,$$

as the integral whose absolute value measures the amplitude of the disturbance. Let us suppose that the aperture is the same sort of rectangle considered above, extending from $-a/2$ to $a/2$ along x, from $-b/2$ to $b/2$ along y. Then the integral is

$$\int_{-a/2}^{a/2} e^{2\pi jlx/\lambda} \, dx \int_{-b/2}^{b/2} e^{2\pi jmy/\lambda} \, dy$$
$$= \frac{(e^{\pi jla/\lambda} - e^{-\pi jla/\lambda})}{2\pi jl/\lambda} \frac{(e^{\pi jmb/\lambda} - e^{-\pi jmb/\lambda})}{2\pi jm/\lambda}$$
$$= \frac{\sin (\pi la/\lambda)}{\pi l/\lambda} \frac{\sin (\pi mb/\lambda)}{\pi m/\lambda} \quad (3.1)$$

The intensity is the square of this quantity. Let us consider its dependence on the position of the point P on the screen. The coordinates of this point enter only in the expressions l, m, showing that the pattern increases in size proportionally to the distance, as if it consisted of rays traveling out in straight lines from the small aperture, rather than having an approximately constant size as with the Fresnel diffraction (see Fig. 32). When we consider the detailed behavior of the intensity as a function of the angle, we find that this can be written as $a^2 \dfrac{\sin^2 (\pi l a / \lambda)}{(\pi l a / \lambda)^2}$ times a similar function of m, giving a curve of the form $\dfrac{\sin^2 \alpha}{\alpha^2}$, where $\alpha = \dfrac{\pi l a}{\lambda}$. This function becomes unity when $\alpha = 0$, goes to zero for $\alpha = \pi, 2\pi, 3\pi, \cdots$, with maxima of intensity approximately midway between. The maxima decrease rapidly in intensity. Thus at the points $3\pi/2, 5\pi/2, \ldots$ which are approximately at the second and third maxima, the intensities are only $(2/3\pi)^2$, $(2/5\pi)^2$, $\ldots$, or $0.045, 0.016 \ldots$, compared with the central maximum of 1. Let us see how the size of the fringes depends on the dimensions of the slit. The minima come for $\alpha = n\pi$, or $la/\lambda = n$, $l = n\lambda/a$. Thus we see that the greater the wave length, or the smaller the dimensions of the slit, the larger the pattern becomes.

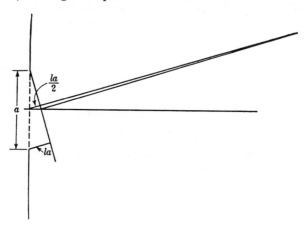

Fig. 35.—Elementary construction for Fraunhofer diffraction.

The positions of the minima can be immediately found by a very elementary argument. Assume for convenience that we are investigating the pattern at a point in the xz plane, so that $m = 0$. Then draw a plane normal to the direction l, passing through one edge of the aperture, as in Fig. 35. This represents a wave front of the diffracted wave, just as it passes one edge of the aperture. From the geometry of the system, this wave front is a distance la from the other edge, or $la/2$ from the middle of the aperture. Now, if the distance of the middle is just a whole number of half wave lengths different from the distance from the edge, the contributions of these two points to the amplitude will just cancel, being just out of phase. The other points of one half of the aperture can all be paired against corresponding points of the other half whose contributions are just out of phase, finally resulting in zero intensity. This situation comes about when $la/2 = n\lambda/2$, where n is an integer, or $l = n\lambda/a$, the same condition found above. Since most of the intensity falls within the first minimum, and since l is the sine of the angle between the ray and the normal to the surface, we may say that by Fraunhofer diffraction the ray is spread out through an angle λ/a.

4. The Circular Aperture, and the Resolving Power of a Lens.— The problem of Fraunhofer diffraction through a circular aperture is slightly more complicated mathematically. Here we must evaluate $\int e^{2\pi i(lx+my)/\lambda}\, da$ over a circle. Let us introduce polar coordinates in the plane of the aperture, so that $x = \rho \cos \theta$, $y = \rho \sin \theta$. Further, on account of symmetry, we may take the point P to be in the xz plane, so that $m = 0$. Then, if ρ_0 is the radius of the aperture, the final result is $\int_0^{2\pi} d\theta \int_0^{\rho_0} e^{2\pi i\rho \cos \theta l/\lambda}\rho\, d\rho$. We integrate with respect to ρ by parts, obtaining for the integral

$$\int_0^{2\pi} d\theta \left[\frac{\rho_0 e^{2\pi i\rho_0 \cos \theta l/\lambda}}{2\pi j \cos \theta l/\lambda} - \frac{(e^{2\pi i\rho_0 \cos \theta l/\lambda} - 1)}{(2\pi j \cos \theta l/\lambda)^2} \right].$$

For the integration with respect to θ, it is necessary to expand the exponentials in series. If we do this, the integrals are in each case integrals of a power of $\cos \theta$, from 0 to 2π. These are easily evaluated, and the result, combining terms, proves to be

$$\pi\rho_0^2 \left[1 - \frac{1}{2}\left(\frac{k}{1}\right)^2 + \frac{1}{3}\left(\frac{k^2}{2!}\right)^2 - \frac{1}{4}\left(\frac{k^3}{3!}\right)^2 + \frac{1}{5}\left(\frac{k^4}{4!}\right)^2 - \cdots \right],$$

where k is an abbreviation for $\pi\rho_0 l/\lambda$. If we recall the formulas for Bessel's functions, discussed in Appendix VII, we can see without difficulty that this is equal to $(\rho_0\lambda/l)J_1[2\pi\rho_0(l/\lambda)]$. It is not hard, using some of the properties of Bessel's functions, to prove this formula directly, without the use of series.

From the series, we see that the intensity has a maximum for $l = 0$, the center of the pattern. As l increases, we can see the behavior most easily from the expression in terms of Bessel's functions. Since J_1 has an infinite number of zeros, there are an infinite number of light and dark fringes. The first dark band comes at the first zero of J_1, which from tables is at $2\pi\rho_0 l/\lambda = 1.2197\pi$, $\rho_0 l/\lambda = 0.61$. The next is at $\rho_0 l/\lambda = 1.16$, and so on, with maxima between. We see that, except for a numerical factor, the pattern from a circular aperture has about the same dimensions as that from a square aperture. Thus if the side of the square were equal to the diameter of the circle, $2\rho_0$, the first dark fringe would be at $2\rho_0 l/\lambda = 1$, $\rho_0 l/\lambda = 0.5$, and the next one at 1.0.

Whenever light passes through a lens, it is not only refracted, but it has passed through a circular aperture, the size of the lens itself or of the diaphragm that stops it down, and as a result it is diffracted. Suppose, for example, that the lens is the objective of a telescope, and that parallel light falls on it, as from an infinitely small or distant star. Then after passing through the diaphragm, the light will no longer be a plane wave, but will have intensity in different directions, as shown in the last paragraph. The central maximum will have an angular radius of $0.61\lambda/\rho_0$, where ρ_0 is now the radius of the telescope objective. The resulting waves are just as if the light came from an object of this size, but passed through no diaphragm. When the telescope focuses the radiation, the result will be not a single point of light, but a circular spot surrounded by fringes, as of a star of finite diameter. For this reason, the telescope is not a perfect instrument, and one would say that its resolving power was only enough to resolve the angle $0.61\lambda/\rho_0$. This is usually taken to mean the following: If two stars had an actual angular separation of this amount, the center of the image of one star would lie on the first dark fringe of the other, and the patterns would run into each other so that they could be just resolved. We see that the larger the aperture of the telescope, or the smaller the wave length, the better is the resolution. The same general situation holds for microscope lenses.

5. Diffraction from Several Slits; the Diffraction Grating.—Suppose we have a number N of equal, parallel slits, equally spaced. Let each have a width a along the x axis, and let the spacing on centers be d, so that the centers come at $x = 0, d, \cdots (N-1)d$. Now let us find the Fraunhofer pattern. The part of the integral depending on y will be just as with the single slit, and we leave it out of account. We are left with

$$\int_{-a/2}^{a/2} e^{2\pi jlx/\lambda}\, dx + \int_{d-a/2}^{d+a/2} e^{2\pi jlx/\lambda}\, dx + \cdots + \int_{(N-1)d-a/2}^{(N-1)d+a/2} e^{2\pi jlx/\lambda}\, dx$$

But this is, as we can immediately see, simply

$$\int_{-a/2}^{a/2} e^{2\pi jlx/\lambda}\, dx\, [1 + e^{2\pi jld/\lambda} + e^{2\pi jl2d/\lambda} + \cdots + e^{2\pi jl(N-1)d/\lambda}].$$

By the formula for the sum of a geometric series, this is

$$\int_{-a/2}^{a/2} e^{2\pi jlx/\lambda}\, dx \left(\frac{1 - e^{2\pi jlNd/\lambda}}{1 - e^{2\pi jld/\lambda}} \right).$$

Let the first term be A, the amplitude due to a single slit, which we have already evaluated. Now to find the intensity we multiply this by its conjugate, which gives

$$A^2 \frac{1 - \cos\,(2\pi lNd/\lambda)}{1 - \cos\,(2\pi ld/\lambda)} = A^2 \frac{\sin^2\,(\pi lNd/\lambda)}{\sin^2\,(\pi ld/\lambda)}. \tag{5.1}$$

That is, with N slits the actual intensity is that with one slit, but multiplied by a certain factor. This factor goes through zero when lNd/λ is an integer, so that l equals an integer multiplied by λ/Nd. This gives fringes with a narrow spacing, characteristic of the whole distance Nd occupied by the set of apertures, modulating the other pattern, and they are what are usually called "interference fringes," since they are due, not to diffraction from a single aperture, but to interference between different apertures. But, in addition to this, the denominator results in having these fringes of different heights. The minimum height occurs when the denominator equals unity, when the fringes are of height A^2, and the most intense fringes come when the denominator is zero. Here the ratio of numerator to denominator is evidently finite, and gives fringes of height N^2A^2. Thus the greater N is, the greater the disparity in height between the largest and smallest maximum. Evidently every Nth maximum will be high, and the high ones will be spaced according to the law $ld/\lambda = k$, an integer.

Now suppose N becomes very great, as in a diffraction grating. Then the small maxima will become so weak compared with the strong ones that only the latter need be considered. The latter will seem to consist of a set of sharp lines, with darkness between. These sharp lines come, as we have seen, at angles θ to the normal given by $k\lambda = d \sin \theta$, where k is an integer, and $\sin \theta = l$. This is the ordinary diffraction grating formula, where k is 0 for the central image, 1 for the first-order spectrum, 2 for the second order, and so on. But we cannot entirely neglect the fact that there are other small maxima near the important ones. Thus for $ld/\lambda = k$, the intensity is N^2A^2. This comes for $lNd/\lambda = Nk$. But for $lNd/\lambda = Nk + \frac{3}{2}$, we again have a secondary maximum, whose height is now

$$\frac{A^2}{\sin^2 (\pi ld/\lambda)} = \frac{A^2}{\sin^2 \pi[(Nk + \frac{3}{2})/N]} = \frac{A^2}{\sin^2 \pi[k + 3/(2N)]}.$$

Now $\sin^2 \left(\pi k + \dfrac{3\pi}{2N}\right) = \left(\dfrac{3\pi}{2N}\right)^2$ approximately, if N is large, so that the height of the maximum is $4N^2A^2/9\pi^2$, or about 0.045 of the height of the highest maximum. Thus the first few secondary maxima cannot be neglected.

To get an idea of the width of the region through which the intensity is considerable, we may take the width of the first maximum. From the center to the first dark fringe, this is given by the fact that at the center $lNd/\lambda = Nk$, at the dark fringe $Nk + 1$, so that $\Delta l = \lambda/Nd$. This is closely connected with the resolving power of a grating. For a single frequency gives not a sharp set of lines, one for each order, but a set broadened by the amount we have found. Thus two neighboring frequencies, differing by $\Delta\lambda$, could not be resolved if the first minimum of one lay opposite the maximum of the other. Since $l = \lambda k/d$, this would be the case if $\Delta l = \Delta\lambda k/d = \lambda/Nd$, or if $\Delta\lambda/\lambda = 1/Nk$. The resolving power thus increases as the number of lines in the grating increases, and as the order of the spectrum increases.

Problems

1. Carry through a discussion of Fresnel diffraction from a slit, when the source is at a finite distance, directly behind the center of the slit. In what ways will the result differ from the case we have discussed?

2. Light of wave length 6,000 A falls in a parallel beam on a slit 0.1 mm broad. Work out numerical values for the intensity distribution across the slit, at three distances, first in which the Fresnel fringes are small compared with the size of the pattern, second in which they are of the same order of magnitude, and third in which they are Fraunhofer fringes. Either construct Cornu's spiral yourself, from tables of Fresnel's integrals, or use the one of Fig. 34.

3. Find the coordinates of the points at which Cornu's spiral winds up on itself. From the chord between these points, compute the intensity behind an infinitely broad slit, which essentially means no slit at all. Find whether this agrees with what you should expect it to be.

4. Prove that the maxima of the function

$$\frac{\sin^2 (\pi la/\lambda)}{(\pi la/\lambda)^2} = \frac{\sin^2 \alpha}{\alpha^2}$$

are determined by the equation $\alpha = \tan \alpha$. Find the first three solutions of this transcendental equation, and compare them with the approximate solutions $\alpha = 3\pi/2,\ 5\pi/2,\ 7\pi/2$.

5. Discuss the Fresnel diffraction pattern caused by an edge coincident with the y axis, the screen occupying one-half the xy plane. The diffraction pattern is obtained in a plane parallel to the xy plane and a distance R from it. Plot the variation of intensity of light along the x direction from a region inside the shadow to well into the directly illuminated area. Prove that the intensity of light just at the edge of the geometrical shadow is one-fourth of its value if there were no diffraction edge.

6. Evaluate the Fresnel integrals $\int_0^u \cos \dfrac{\pi}{2} u^2\, du$ and $\int_0^u \sin \dfrac{\pi}{2} u^2\, du$ in a power series. What is the range of convergence of these series?

7. Consider the Fresnel integrals in the form $\int_0^x \cos x^2\, dx$ and $\int_0^x \sin x^2\, dx$. Integrate each of these by parts according to the scheme

$$\int_0^x \cos x^2\, dx = x \cos x^2 + 2 \int_0^x x^2 \sin x^2\, dx$$

and continue this process. Show that one obtains series for the above integrals of the form

$$\int_0^x \cos x^2\, dx = S_1 \cos x^2 + S_2 \sin x^2$$

and

$$\int_0^x \sin x^2\, dx = S_1 \sin x^2 - S_2 \cos x^2$$

where S_1 and S_2 are power series in x. Find these series and their range of convergence.

8. Find a semiconvergent series for the Fresnel integrals by the following method: Write $\int_x^\infty \cos x^2\, dx = \int_x^\infty \dfrac{\cos x^2}{x} x\, dx$, and integrate by parts, repeating the process. Then write the results in the form

$$\int_x^\infty \cos x^2\, dx = \sigma_1 \cos x^2 - \sigma_2 \sin x^2 + R_c$$

where σ_1 and σ_2 are finite series of n terms each in inverse powers of x. R_c is the remainder. Show that it is given by

$$R_c = (-1)^n \frac{1 \cdot 3 \cdot 5 \cdots (4n-1)}{2^{2n}} \int_x^\infty \frac{\cos x^2\, dx}{x^{4n}}.$$

Note that the series σ_1 and σ_2 are alternating series with terms that initially decrease with increasing n but eventually increase again. Thus the best approximation will be obtained when one stops at such a value of n that the ratio of the remainder when $n + 1$ terms are retained to that for n terms is most nearly equal to unity.

Since

$$|R_c| < R_n = \frac{1 \cdot 3 \cdot 5 \cdots (4n-1)}{2^{2n}} \int_x^\infty \frac{dx}{x^{4n}}$$

evaluate R_{n+1}/R_n, and show that the best approximation occurs for n equal about to $x^2/2$.   Find the corresponding series for $\int_x^\infty \sin x^2 \, dx$.

9. Show that in a diffraction grating all the even-order spectra will be missing if the slit separation d is twice the slit width a.   Which orders will be missing if $d = 3a$?   Plot the intensity distribution in the diffraction pattern formed by a grating of four equally spaced slits with $d = 3a$.

APPENDIX I

VECTORS

The study of vectors and vector operations involves two branches: vector algebra, including the addition and multiplication of vectors, and the relations between the components of vectors in different coordinate systems; and vector analysis, including the differential vector operations, corresponding integral relations, and the general theorems concerning integrals. We shall treat both subdivisions in the present appendix.

Vectors and Their Components.—We shall denote a vector, a quantity having direction as well as magnitude, by bold-faced type, as **F**. Vectors are often described by giving their components along three axes at right angles, as F_x, F_y, F_z. Their mathematical relationships are conveniently stated in terms of their components. Thus their law of addition is the parallelogram law; in terms of components, this means that, if two vectors **F** and **G** have components F_x, F_y, F_z, and G_x, G_y, G_z, respectively, the components of the sum **F** + **G** are $(F_x + G_x)$, $(F_y + G_y)$, $(F_z + G_z)$, as we show graphically in Fig. 36. To multiply a vector by a constant, as C, the vector must be increased in length by the factor C, leaving its direction unchanged; this amounts to multiplying each component by the constant C, so that the components of C**F** are CF_x, CF_y, CF_z. Often a constant like C is called a

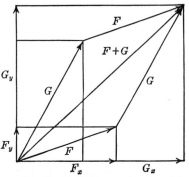

FIG. 36.—Parallelogram law for addition of vectors. The vector **F** + **G**, the diagonal of the parallelogram of sides **F**, **G**, is the vector sum of **F** and **G**. Its x component is $F_x + G_x$, its y component $F_y + G_y$.

"scalar," to distinguish it from a vector. A scalar is a quantity that has magnitude but not direction, a vector having both magnitude and direction.

It is often useful to write vectors in terms of three so-called "unit vectors," **i, j, k.** Here **i** is a vector of unit length, pointing along the

x axis, and similarly **j** has unit length and points along the y axis, and **k** along the z axis.   Now we can build up a vector **F** out of them, by forming the quantity $\mathbf{i}F_x + \mathbf{j}F_y + \mathbf{k}F_z$.   This is the sum of three vectors, one along each of the three axes; and the first, which is just the component of the whole vector along the x axis, is F_x, and the other components likewise are F_y and F_z.   Thus the final vector has the components F_x, F_y, F_z, and is just the vector **F**.

By the magnitude of a vector we mean its length.   By the three-dimensional analogue to the Pythagorean theorem, by which the square of the diagonal of a rectangular prism is the sum of the squares of the three sides, the magnitude of the vector **F** equals $\sqrt{F_x^2 + F_y^2 + F_z^2}$. We often speak of unit vectors, vectors whose magnitude is 1.   The component of a vector in a given direction is simply the projection of the vector along a line in that direction.   It evidently equals the magnitude of the vector, times the cosine of the angle between the direction of the vector and the desired direction.   As a special example, the component of a vector **F** along the x axis is F_x, and this must equal the magnitude of **F**, times the cosine of the angle between **F** and x.   If this angle is called (F,x), then we must have

$$\cos (F,x) = \frac{F_x}{\sqrt{F_x^2 + F_y^2 + F_z^2}}$$

with similar formulas for y and z components.   The three cosines of the angles between a given direction, as the direction of the vector **F,** and the three axes, are called "direction cosines," and are often denoted by letters l, m, n, so that in this case we have $l = \cos (F,x)$, etc.   It follows immediately that $l^2 + m^2 + n^2 = 1$.   We can give a simple interpretation of the direction cosines of any direction: they are the components of a unit vector in the desired direction, along the three coordinate axes.

Scalar and Vector Products of Two Vectors.—Multiplication of two vectors is a somewhat arbitrary process, governed by rules that we must postulate.   It has proved to be convenient to define two entirely independent products, called the "scalar product" and the "vector product."   We shall first consider the scalar product.   The scalar product of two vectors **F** and **G** is denoted by **F · G,** and by definition it is a scalar, equal to either (1) the magnitude of **F** times the magnitude of **G** times the cosine of the angle between; or (2) the magnitude of **F** times the projection of **G** on **F**; or (3) the magnitude of **G** times the projection of **F** on **G**.   From the preceding section we see that these definitions are equivalent.

It is often useful to have the scalar product of two vectors in terms of the components along x, y, and z. We find this by writing in terms of $\mathbf{i}$, $\mathbf{j}$, and $\mathbf{k}$. Thus we have

$$
\begin{aligned}
\mathbf{F} \cdot \mathbf{G} &= (\mathbf{i}F_x + \mathbf{j}F_y + \mathbf{k}F_z) \cdot (\mathbf{i}G_x + \mathbf{j}G_y + \mathbf{k}G_z) \\
&= (\mathbf{i} \cdot \mathbf{i})F_xG_x + (\mathbf{i} \cdot \mathbf{j})F_xG_y + (\mathbf{i} \cdot \mathbf{k})F_xG_z + (\mathbf{j} \cdot \mathbf{i})F_yG_x \\
&\quad + (\mathbf{j} \cdot \mathbf{j})F_yG_y + (\mathbf{j} \cdot \mathbf{k})F_yG_z + (\mathbf{k} \cdot \mathbf{i})F_zG_x + (\mathbf{k} \cdot \mathbf{j})F_zG_y + (\mathbf{k} \cdot \mathbf{k})F_zG_z.
\end{aligned}
$$

But, by the fundamental definition,

$$
\begin{aligned}
\mathbf{i} \cdot \mathbf{i} &= \mathbf{j} \cdot \mathbf{j} = \mathbf{k} \cdot \mathbf{k} = 1, \\
\mathbf{i} \cdot \mathbf{j} &= \mathbf{j} \cdot \mathbf{i} = \mathbf{j} \cdot \mathbf{k} = \mathbf{k} \cdot \mathbf{j} = \mathbf{k} \cdot \mathbf{i} = \mathbf{i} \cdot \mathbf{k} = 0.
\end{aligned} \tag{1}
$$

Thus

$$
\mathbf{F} \cdot \mathbf{G} = F_xG_x + F_yG_y + F_zG_z. \tag{2}
$$

The scalar product has many uses, principally in cases where we are interested in the projection of vectors. For example, the scalar product of a vector with a unit vector in a given direction equals the projection of the vector in the desired direction. The scalar product of a vector with itself equals the square of its magnitude, and is often denoted by F^2. The scalar product of two unit vectors gives the cosine of the angle between the directions of the two vectors. To prove that two vectors are at right angles, we need merely prove that their scalar product vanishes.

The vector product of two vectors $\mathbf{F}$ and $\mathbf{G}$ is denoted by $(\mathbf{F} \times \mathbf{G})$, and by definition it is a vector, at right angles to the plane of the two vectors, equal in magnitude to either (1) the magnitude of $\mathbf{F}$ times the magnitude of $\mathbf{G}$ times the sine of the angle between them; or (2) the magnitude of $\mathbf{F}$ times the projection of $\mathbf{G}$ on the plane normal to $\mathbf{F}$; or (3) the magnitude of $\mathbf{G}$ times the projection of $\mathbf{F}$ on the plane normal to $\mathbf{G}$. We must further specify the sense of the vector, whether it points up or down from the plane. This is shown in Fig. 37, where we see that $\mathbf{F}$, $\mathbf{G}$, and $\mathbf{F} \times \mathbf{G}$ have the same relations as the coordinates x, y, z in a right-handed system of coordinates.

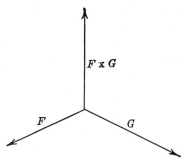

Fig. 37.—Direction of the vector product.

Another way to describe the rule in words is that, if one rotates $\mathbf{F}$ into $\mathbf{G}$, the rotation is such that a right-handed screw turning in that direction would be driven along the direction of the vector product. From this rule, we note one interesting fact: if we interchange the order of the factors, we reverse the vector. Thus

$$(\mathbf{F} \times \mathbf{G}) = -(\mathbf{G} \times \mathbf{F}).$$

We can compute the vector product in terms of the components, much as we did with the scalar product. Thus we have

$$\begin{aligned}
\mathbf{F} \times \mathbf{G} = (\mathbf{i} \times \mathbf{i})F_xG_x + (\mathbf{i} \times \mathbf{j})F_xG_y + (\mathbf{i} \times \mathbf{k})F_xG_z \\
+ (\mathbf{j} \times \mathbf{i})F_yG_x + (\mathbf{j} \times \mathbf{j})F_yG_y + (\mathbf{j} \times \mathbf{k})F_yG_z \\
+ (\mathbf{k} \times \mathbf{i})F_zG_x + (\mathbf{k} \times \mathbf{j})F_zG_y + (\mathbf{k} \times \mathbf{k})F_zG_z.
\end{aligned}$$

But now, as we readily see from the definition,

$$(\mathbf{i} \times \mathbf{i}) = (\mathbf{j} \times \mathbf{j}) = (\mathbf{k} \times \mathbf{k}) = 0$$

(as, in fact, the vector product of any vector with itself is zero), and

$$(\mathbf{i} \times \mathbf{j}) = -(\mathbf{j} \times \mathbf{i}) = \mathbf{k}, \qquad (\mathbf{j} \times \mathbf{k}) = -(\mathbf{k} \times \mathbf{j}) = \mathbf{i},$$
$$(\mathbf{k} \times \mathbf{i}) = -(\mathbf{i} \times \mathbf{k}) = \mathbf{j}. \quad (3)$$

Hence, rearranging terms, we have

$$\mathbf{F} \times \mathbf{G} = \mathbf{i}(F_yG_z - F_zG_y) + \mathbf{j}(F_zG_x - F_xG_z) + \mathbf{k}(F_xG_y - F_yG_x). \quad (4)$$

The Differentiation of Vectors.—We have seen that there are at least three processes of multiplication involving vectors: the multiplication of a vector by a scalar, the scalar product of two vectors, and the vector product of two vectors. In a somewhat similar way there are a number of differential operations involving vectors, all with their special uses. The simplest of these is the differentiation of a vector with respect to a scalar, such as the time, in which the x, y, and z components of the time derivative of a vector are simply the time derivative of the x, y, and z components of the vector.

The next type of differentiation that we consider is the differentiation of a scalar with respect to x, y, and z, to give a vector, the gradient, which we encountered in Chap. I, Sec. 3, in discussing the relationship between the electric field $\mathbf{E}$ and the scalar potential. The gradient of a potential φ is defined as

$$\operatorname{grad} \varphi = \left(\mathbf{i}\,\frac{\partial}{\partial x} + \mathbf{j}\,\frac{\partial}{\partial y} + \mathbf{k}\,\frac{\partial}{\partial z}\right)\varphi.$$

Here we have written the gradient symbolically as the product of a vector operator, and the scalar φ. This vector operator is ordinarily denoted by a special symbol ∇ (pronounced "del"), defined by

$$\nabla = \mathbf{i} \frac{\partial}{\partial x} + \mathbf{j} \frac{\partial}{\partial y} + \mathbf{k} \frac{\partial}{\partial z}. \tag{5}$$

Whenever we use the operator ∇, we understand that the differentiations are to operate on whatever appears to the right of the operator. With this definition, we see that we may write the identity

$$\text{grad } \varphi \equiv \nabla\varphi.$$

In some texts on vectors, the gradient is simply denoted by $\nabla\varphi$. As we see in Chap. I, Sec. 3, the physical significance of the gradient is simple. The gradient of a scalar points at right angles to the surfaces on which that function is constant, or in the direction in which the function changes most rapidly with position, and its magnitude measures the rate of change of the function in that direction. Its component in any direction, often called the "directional derivative" of the function in that direction, measures the rate of change of the function in that direction, as for instance $\partial\varphi/\partial x$ measures the rate of change of φ with x.

Using the vector operator ∇, we can now define two types of derivatives of a vector: if we have a vector $\mathbf{F}$, we can define derivatives by taking the scalar or the vector product of ∇ with the vector, resulting in $\nabla \cdot \mathbf{F}$ and $\nabla \times \mathbf{F}$. These are the quantities defined as the divergence and the curl. From the definition (5), we have

$$\text{div } \mathbf{F} = \nabla \cdot \mathbf{F} = \frac{\partial F_x}{\partial x} + \frac{\partial F_y}{\partial y} + \frac{\partial F_z}{\partial z}$$

and

$$\text{curl } \mathbf{F} = \nabla \times \mathbf{F} = \mathbf{i}\left(\frac{\partial F_z}{\partial y} - \frac{\partial F_y}{\partial z}\right) + \mathbf{j}\left(\frac{\partial F_x}{\partial z} - \frac{\partial F_z}{\partial x}\right) + \mathbf{k}\left(\frac{\partial F_y}{\partial x} - \frac{\partial F_x}{\partial y}\right).$$

There are also various vector operations involving second derivatives. The most familiar ones involve the operator $\nabla \cdot \nabla = \nabla^2$, which written out is

$$\nabla^2 = \frac{\partial^2}{\partial x^2} + \frac{\partial^2}{\partial y^2} + \frac{\partial^2}{\partial z^2}.$$

This is the operator that we encounter often in the wave equation, Laplace's equation, etc., and that is often called the "Laplacian," for that reason. Being a scalar operator, it can be applied to either a scalar or a vector, and both forms frequently occur. Another second differential operator that is encountered often in electromagnetic theory, though not often in mechanics, is the vector operator curl curl, applied to a vector: curl curl $\mathbf{F} = \nabla \times (\nabla \times \mathbf{F})$. We prove in a problem the useful relation

$$\text{curl curl } \mathbf{F} = \text{grad div } \mathbf{F} - \nabla^2\mathbf{F},$$

reducing this operator to the Laplacian, which we have already met, and to grad div $\mathbf{F} = \nabla(\nabla \cdot \mathbf{F})$. This completes the list of the vector operations that are often encountered.

The Divergence Theorem and Stokes's Theorem.—There are two vector theorems involving integrals, somewhat similar in principle to the theorem regarding integration by parts in calculus, which are of great importance in vector analysis. These are the divergence theorem, sometimes known as Gauss's theorem, and Stokes's theorem. We shall now prove these theorems. The divergence theorem relates to a closed volume V in space, and the surface S that bounds it. It is assumed that there is a vector function of position $\mathbf{F}$. The theorem then states that the surface integral of the normal component of $\mathbf{F}$, over the surface S, equals the volume integral of div $\mathbf{F}$, over the volume V. That is,

$$\int\int_S F_n \, da = \int\int_S \mathbf{F} \cdot \mathbf{n} \, da = \int\int\int_V \text{div } \mathbf{F} \, dv, \qquad (6)$$

where $\mathbf{n}$ is unit vector along the outer normal, so that $\mathbf{F} \cdot \mathbf{n}$ is another way of writing F_n, the component of $\mathbf{F}$ along the outer normal. To prove our theorem, we start by dividing up the volume V into thin

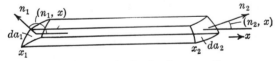

Fig. 38.—Construction for the divergence theorem.

elements bounded by planes $y = $ constant, $z = $ constant, as in Fig. 38. The component F_x will be a function of x along such an element, and we have obviously

$$\int_{x_1}^{x_2} \frac{\partial F_x}{\partial x} \, dx = F_x \Big|_{x_1}^{x_2}, \qquad (7)$$

where x_1, x_2 are the values of x at which the element cuts through the surface S. Let $\mathbf{n}_1$, $\mathbf{n}_2$ be the outer normals at these two ends of the element, and let da_1, da_2 be the corresponding areas of surface bounding the ends of the element. We shall have

$$-da_1 \cos (n_1,x) = dy\, dz$$
$$da_2 \cos (n_2,x) = dy\, dz$$

where (n_1,x) and (n_2,x) refer to the angles between the vectors $\mathbf{n}_1$ and $\mathbf{n}_2$ and the x axis, where the negative sign in the first equation arises because the outer normal has a negative projection on the x axis at the end x_1, and where dy, dz represent the thickness of the element along the y and z directions.

Multiplying both sides of (7) by $dy\, dz$, we then have

$$\int_{x_1}^{x_2} \frac{\partial F_x}{\partial x}\, dx\, dy\, dz = [F_{x1} \cos (n_1,x)\, da_1 + F_{x2} \cos (n_2,x)\, da_2].$$

If we carry out a summation over all elements of this type, the integral on the left will become a volume integral over V, the sum on the right will become a surface integral over S, and we have

$$\int \int_V \int \frac{\partial F_x}{\partial x}\, dv = \int \int_S F_x \cos (n,x)\, da. \tag{8}$$

We now proceed similarly with y and z, breaking up the volume into thin elements with their axes along y and z, and obtain two other equations similar to (8). Adding them, we have

$$\int \int_V \int \left(\frac{\partial F_x}{\partial x} + \frac{\partial F_y}{\partial y} + \frac{\partial F_z}{\partial z} \right) dv$$
$$= \int \int_S [F_x \cos (n,x) + F_y \cos (n,y) + F_z \cos (n,z)]\, da.$$

But the integrand on the right is just the scalar product $\mathbf{F} \cdot \mathbf{n}$, so that we have proved our theorem (6). This theorem, the divergence theorem, is met in Poisson's equation, the equation of continuity, and other places, and forms the basis of Green's theorem, which is proved in Sec. 4, Chap. II.

A simplified version of our proof, though not entirely satisfactory, is instructive. This comes from proving the theorem for a volume consisting of an infinitesimal rectangular parallelepiped, bounded by $x, x + dx, y, y + dy, z, z + dz$. First we compute the surface integral of F_n over the six faces of the volume. For the face at x, F_n is $-F_x$, and the integral is $-F_x(x)\, dy\, dz$, where the value of F_x is to be computed at x. For the face at $x + dx$, F_n is F_x, and the integral is $F_x(x + dx)\, dy\, dz$. Thus the surface integral over these two faces is approximately $(\partial F_x/\partial x)\, dx\, dy\, dz$. Adding similar contributions from the faces normal to the y and z axes, we find for the total surface integral the amount div $\mathbf{F}\, dx\, dy\, dz$, which is what the divergence theorem would give. This treatment suggests a simple physical definition for the divergence of a vector: it is the total outward flux of the vector per unit volume. To get from this form of the theorem to that involving a finite volume, we may subdivide the finite volume into infinitesimal parallelepipeds. The total flux outward of the vector through the surface of the finite volume is just equal to the sum of the fluxes outward over the surfaces of the infinitesimal volumes; for at each interior surface of separation between infinitesimal volumes, the flux outward from one volume is just balanced by the flux into its neighbor, leaving only the contributions of the exterior surfaces. The weakness of this proof lies only in the fact that a real volume cannot be built up entirely of infinitesimal parallelepipeds; around the boundary there would have to be infinitesimal volumes with surfaces inclined to the coordinate planes, which this treatment does not consider. Our earlier treatment, however, removes this objection.

Next we consider Stokes's theorem. This theorem relates to a closed line L in space, and a surface S that is bounded by L. Again we have a vector function of position $\mathbf{F}$. Stokes's theorem states that the line integral of the tangential component of $\mathbf{F}$, around L, equals the surface integral of the normal component of curl $\mathbf{F}$, over S. That is,

$$\int_{L} \mathbf{F} \cdot \mathbf{ds} = \int \int_{S} (\text{curl } \mathbf{F}) \cdot \mathbf{n}\, da, \tag{9}$$

where $\mathbf{ds}$ is a vector element of distance around the boundary, and $\mathbf{n}$ is the normal to the surface, chosen to point up if the positive direction of traversing the boundary is chosen (that is, if we go around it in a counterclockwise direction). We shall give a simplified discussion of this theorem, similar to the second method we used for handling the divergence theorem. Let us subdivide the surface S into a set of small approximately rectangular areas. We shall prove the theorem for each of the separate areas, and then combine them. As we see from Fig. 39, the contributions to the line integrals from the

FIG. 39.—Surface for discussing Stokes's theorem.

interior boundaries will cancel, so that the sum of all the line integrals will equal the line integral around the perimeter. The sum of the surface integrals over the separate rectangles will clearly equal the surface integral over the whole surface. Thus by adding the separate contributions we arrive at the theorem (9), which we wish to prove. The only reservation is that the line integral is computed for a saw-tooth type of curve approximating the actual boundary. It is not hard to show, though we shall omit it, that the line integral $\int \mathbf{F} \cdot \mathbf{ds}$ over a saw-tooth curve approximating the actual curve sufficiently smoothly differs by a negligible amount from the integral over the actual curve, so that we can construct a rigorous proof of the theorem without trouble.

We must then prove our theorem (9) for a small rectangular area. Let us choose the x and y axes to point along the sides of the rectangle, the z axis along the normal $\mathbf{n}$; if we can prove the result in this coordinate system, it will have to hold in any other coordinate system as well. The rectangle is considered to be bounded by the values x, $x + dx$, y, $y + dy$, as in Fig. 40. The surface integral on the right of (9) is then

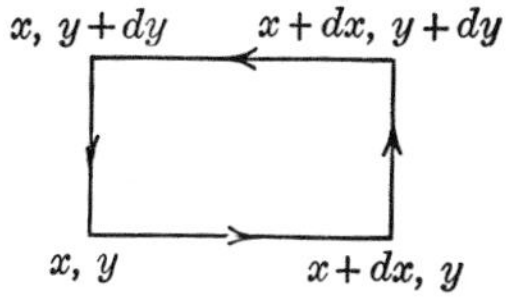

FIG. 40.—Circuit for proving Stokes's theorem.

$(\partial F_y/\partial x - \partial F_x/\partial y)\, dx\, dy$. Let us next compute $\int \mathbf{F} \cdot \mathbf{ds}$ for the element of area. It is evidently

$$F_x(x,y)\, dx + F_y(x + dx,y)\, dy - F_x(x,y + dy)\, dx - F_y(x,y)\, dy$$
$$= \left(\frac{\partial F_y}{\partial x} - \frac{\partial F_x}{\partial y} \right) dx\, dy,$$

if we go around so as always to keep the surface on the left. Thus the theorem (9) is true for the infinitesimal rectangular surface, and by the argument above it then holds for a finite surface as well.

The physical meaning of the curl is easily seen from Stokes's theorem: it is the line integral of the tangential component of the vector around an area, per unit area. A vector field has a curl when its lines of force close on themselves, like the magnetic lines around a wire carrying an electric current, or like the lines of flow in a whirlpool-like motion of a fluid. A simple example, which makes the significance of the curl very clear, is the vector velocity associated with the rigid rotation of a body. Let a body have an angular velocity ω about the z axis. Then the linear velocity at point x, y, z is given by $v_x = -\omega y$, $v_y = \omega x$, $v_z = 0$. We compute the curl, and find at once that it is along the z direction, has a constant value independent of position, and is of magnitude 2ω. Thus it is simply twice the angular velocity vector of the rotating motion.

Problems

1. Find the angle between the diagonal of a cube and one of the edges.   (*Hint:* Regard the diagonal as a vector $i + j + k$.)

2. Given a vector $i + 2j + 3k$, and a second $i - 2j + ak$, find a so that the two vectors are at right angles to each other.

3. Prove that $lx + my + nz = k$, where l, m, n, k are constants, and $l^2 + m^2 + n^2 = 1$ is the equation of a plane whose normal has the direction cosines l, m, n, and whose shortest distance from the origin is k.

4. Prove that $A \cdot (B \times C) = B \cdot (C \times A) = C \cdot (A \times B)$, where A, B, C are any vectors.   Show that these are equal to the determinant

$$\begin{vmatrix} A_x & A_y & A_z \\ B_x & B_y & B_z \\ C_x & C_y & C_z \end{vmatrix}.$$

5. Prove that $A \times (B \times C) = B(A \cdot C) - C(A \cdot B)$, where A, B, C are any vectors.

6. Prove that $\operatorname{div} aF = a \operatorname{div} F + (F \cdot \operatorname{grad} a)$, where a is a scalar, F a vector.

7. Prove that $\operatorname{curl} aF = a \operatorname{curl} F + [(\operatorname{grad} a) \times F]$, where a is a scalar, F a vector.

8. Prove that $\operatorname{div} (F \times G) = (G \cdot \operatorname{curl} F) - (F \cdot \operatorname{curl} G)$, where F, G are vectors.

9. Prove that $\operatorname{div} \operatorname{curl} F = 0$, where F is any vector.

10. Prove that $\operatorname{curl} \operatorname{curl} F = \operatorname{grad} \operatorname{div} F - \nabla^2 F$, where F is any vector.

11. Prove that $\operatorname{curl} \operatorname{grad} a = 0$, where a is any scalar.

APPENDIX II

UNITS

One of the complications of electromagnetic theory is the fact that a number of different systems of units are in common use. There are two so-called "absolute" cgs systems, both based on the centimeter as unit of distance, gram as unit of mass, and second as unit of time. These are the electrostatic and electromagnetic units, respectively (often abbreviated esu and emu). There is a combination of these, in which certain quantities are denoted in the electrostatic and certain others in the electromagnetic units, which is called the "Gaussian," or "mixed," units. There are the so-called "practical units," based largely on the electromagnetic units. There is the system used in this volume, the so-called mks system, in which, by using practical units for the electromagnetic quantities, but by using meters as units of distance, and kilograms as units of mass, we convert the practical system into an absolute system (that is, into a system in which various numerical constants can be set equal to unity, so that they drop out of the equations). Finally, in all these systems, we have a choice of so-called "rationalized" and "unrationalized" units, differing by factors of 4π, in which the unrationalized units give simpler formulas when we are dealing with problems of spherical symmetry, as the force between two point charges (where the force field is spherically symmetrical), while the rationalized units give simpler formulas when we are dealing with problems with rectangular symmetry, as the capacity of a parallel-plate condenser, or the properties of a plane wave. Of all these systems, the unrationalized Gaussian units, and the rationalized mks units, have the widest use and are the most valuable. In this appendix we take up the relations among these units, and the definitions of all the types of units.

We start with the unrationalized electrostatic units. Here the fundamental definition is that of the unit charge. It is so chosen as to eliminate the constant in Coulomb's law, which we have stated in mks units in Eq. (2.2), Chap. I. That is, in esu, the force in dynes between two unit charges is given by

$$F = \frac{qq'}{r^2},\tag{1}$$

205

where r is in centimeters; or unit charge is that charge which, placed at a distance of 1 cm from an equal charge, acts on it with a force of 1 dyne. The corresponding rationalized electrostatic units would be defined by the statement $F = qq'/4\pi r^2$, so that unit charge would be that which, placed at a distance of 1 cm from an equal charge, acts on it with a force of $1/4\pi$ dynes. That is, the unit charge in rationalized units is $1/\sqrt{4\pi}$ times as great as in unrationalized units. We shall not, however, consider rationalized esu further.

In unrationalized esu, we define unit current as that in which unit charge passes any point of the circuit per second. We define unit electric field **E** from the force equation, which we have given in Eq. (1.1), Chap. I, for mks units. That is, we assume that

$$\mathbf{F} = q\mathbf{E}, \tag{2}$$

for a static field, so that unit field is that in which unit charge is acted on by a force of 1 dyne. Electrostatic unit of potential is that potential difference which will impart to unit charge an energy of 1 erg. It is clear from this that the field of a charge q, at a distance r, has a magnitude q/r^2, and a scalar potential q/r, in unrationalized esu. The units of **D**, in unrationalized esu, are determined by the arbitrary assumption that $\mathbf{D} = \mathbf{E}$ in empty space, in these units, so that the units of **D** are the same as those of **E**. If now we have a field

$$E = D = \frac{q}{r^2}$$

from a charge q, the outward flux of **D** can be found at once by considering a sphere, and is $4\pi q$. Thus Gauss's law becomes

$$\int \mathbf{D} \cdot \mathbf{n} \, da = 4\pi \sum_i q_i = 4\pi \int \rho \, dv, \tag{3}$$

and the corresponding differential expression is

$$\operatorname{div} \mathbf{D} = 4\pi\rho. \tag{4}$$

With the formulas relating charge and field, we can find the capacity of a condenser in esu, where of course capacity is defined as the charge per unit potential difference. For a parallel-plate condenser of area A sq cm, with a separation of d cm between the plates, and empty space between, the capacity proves to be

$$C = \frac{A}{4\pi d}. \tag{5}$$

We note that the capacity, being an area divided by a length, has the dimension of length in these units; it is sometimes expressed in centimeters. We also note that the factor 4π, which did not appear in Coulomb's law (1) in unrationalized units, has appeared in the formula for the capacity of a parallel-plate condenser, just the reverse of the situation with rationalized units. If now the condenser is filled with a dielectric, the capacity will be κ_e times as great, where κ_e, the dielectric constant, being a dimensionless quantity, is the same in all systems of units. The situation is complicated by the fact, however, that in unrationalized esu the quantity κ_e is often denoted by ϵ, though with a quite different meaning for ϵ from what we have in our mks units. The capacity of a parallel-plate condenser filled with a dielectric, in unrationalized esu, is then

$$C = \frac{\kappa_e A}{4\pi d} \text{ or } \frac{\epsilon A}{4\pi d}. \tag{6}$$

It will be noted that this is related to the value $\epsilon_0 A/d$, given in Sec. 2, Chap. II, for the mks units, by having ϵ in place of ϵ_0 (since the dielectric is not equivalent to free space), and by being divided by 4π. Similarly any of the other expressions for capacities of condensers of various shape, given in that section, can be converted to unrationalized esu by changing ϵ_0 to ϵ, and by dividing by 4π.

A dielectric of course derives its properties from the dipoles within it. The potential of a dipole of moment m (which, of course, is the product of the charge, in esu, times the separation of the opposite charges, in centimeters) is $m \cos \theta/r^2$, instead of the value

$$\frac{m \cos \theta}{4\pi\epsilon_0 r^2}$$

which we have in mks units. We set up a polarization vector $\mathbf{P}$, equal to the dipole moment per unit volume in a dielectric, just as we did in mks units. As before, we find the volume charge density arising from the polarization to be given by the relation $\text{div } \mathbf{P} = -\rho'$. Assuming, as in Chap. IV, that $\mathbf{E}$ is the field arising from all charge, real and polarized, we have $\text{div } \mathbf{E} = 4\pi(\rho + \rho') = 4\pi\rho - 4\pi \text{ div } \mathbf{P}$, $\text{div } (\mathbf{E} + 4\pi\mathbf{P}) = 4\pi\rho$. Thus, comparing with (4), we see that

$$\kappa_e = 1 + \frac{4\pi P}{E} = 1 + 4\pi\chi_e, \tag{7}$$

where χ_e, the susceptibility, in the unrationalized units, is clearly $1/4\pi$ times as great as in rationalized units, as in Eq. (2.3), Chap. IV.

Taking account of the different definition of the dipole moment, we see that, in place of Eq. (1.3), Chap. IX, giving the dielectric constant of a material containing polarizable molecules, we have

$$\kappa_e = 1 + 4\pi \sum_k \frac{N_k e^2/m}{\omega_k^2 - \omega^2 + j\omega g_k}, \tag{8}$$

differing from the formula in mks units by a factor $4\pi\epsilon_0$ in the second term.

We have now discussed most of the electrostatic relations in unrationalized esu. One additional quantity is the conductivity, which is defined by Ohm's law $\mathbf{J} = \sigma\mathbf{E}$, in terms of the esu definitions of current density and electric field. When we come to magnetic quantities, it is less convenient to use esu. We can do so, by defining a magnetic field in terms of the force on a current element, but such magnetic units are seldom used, and we shall not discuss them here.

We next consider the unrationalized electromagnetic units. These are based on the law of force between two current elements as a starting point. Just as the esu of charge is chosen so as to eliminate the constant in Coulomb's law, so the emu of current is chosen so as to eliminate the constant in Eq. (1.4) of Chap. V, giving the force between two current elements. Thus in emu we have for that equation

$$d\mathbf{F} = i_1 i_2 \frac{[\mathbf{ds}_1 \times (\mathbf{ds}_2 \times \mathbf{r})]}{|\mathbf{r}|^3}. \tag{9}$$

That is, unit current in the electromagnetic system is that current which, flowing in 1 cm length of wire, acts on a similar current 1 cm away, appropriately oriented, with a force of 1 dyne. Having defined unit current, we can define unit charge in the electromagnetic system as the charge crossing a given point per second, when unit current is flowing. We can define the emu of potential as the potential difference that will impart to 1 emu of charge an energy of 1 erg. We thus note that we can get definitions of electrostatic quantities in the electromagnetic system of units. It is then a question of experiment (involving in principle the measurement of forces between charges, and forces between currents, and comparing the charges and currents) to find the relation between the electrostatic and electromagnetic units of charge, current, and potential difference. It is found that

$$\begin{aligned}
1 \text{ emu of charge} &= 2.998 \times 10^{10} \text{ esu of charge} \\
1 \text{ emu of current} &= 2.998 \times 10^{10} \text{ esu of current} \\
2.998 \times 10^{10} \text{ emu of potential} &= 1 \text{ esu of potential.} \tag{10}
\end{aligned}$$

The quantity 2.998×10^{10} appearing in the ratio of units is that which proves to express the velocity of light in centimeters per second, and for most purposes it can be approximated by 3×10^{10} cm/sec.

The magnetic induction **B** is determined in the electromagnetic system in such a way that the force equation $\mathbf{F} = q(\mathbf{v} \times \mathbf{B})$ or force per unit volume $= (\mathbf{J} \times \mathbf{B})$ will hold, without additional constants. Thus unit magnetic induction, in emu, is that induction in which a unit current flowing in a wire 1 cm long at right angles to the magnetic field is acted on by a force of 1 dyne. This emu of magnetic induction is almost the only one of the absolute units that is in common use: it is the gauss. We then find that the induction arising from an element of current is

$$d\mathbf{B} = \frac{i \, \mathbf{ds} \times \mathbf{r}}{|\mathbf{r}|^3}, \tag{11}$$

the appropriate form of the Biot-Savart law for emu. Using this law, we can find the value of **B** arising from different forms of stationary currents. Since (11) is related to Eq. (1.3) of Chap. V, the corresponding formula in mks units, by lacking the factor $\mu_0/4\pi$, it is clear that all the formulas for **B** in Secs. 2 and 3 of Chap. V can be converted to emu by multiplication by $4\pi/\mu_0$. Thus the value of **B** at a distance R from an infinite straight wire carrying a current of i (in emu) is $2i/R$, and the value of **B** inside a solenoid with n turns per unit length is $4\pi n i$. The solid angle Ω intercepted by a wire carrying a current of i emu can still be used as a scalar potential, and **B** is given by

$$\mathbf{B} = -i \operatorname{grad} \Omega. \tag{12}$$

If we define the strength of a magnetic dipole as the current, in emu, times the area, the scalar potential, whose negative gradient gives **B,** is $m \cos \theta / r^2$. The vector potential **A** still is related to **B** by the equation $\mathbf{B} = \operatorname{curl} \mathbf{A}$, where now **A** is determined by the equation

$$\mathbf{A} = \int \frac{\mathbf{J}}{r} \, dv, \tag{13}$$

instead of by Eq. (6.4), Chap. V.

From (12), we can see at once that Ampère's law for empty space is

$$\int \mathbf{B} \cdot \mathbf{ds} = 4\pi \Sigma i, \tag{14}$$

or, in differential form,

$$\operatorname{curl} \mathbf{B} = 4\pi \mathbf{J}. \tag{15}$$

In a magnetic medium, however, we have a magnetization current $\mathbf{J}'$

arising from magnetic polarization. Defining magnetic moment as we have just done, we can set up a magnetization vector **M** as the total magnetic polarization per unit volume, and we find, as in Eq. (1.2), Chap. VI, that

$$\mathbf{J}' = \text{curl } \mathbf{M}. \tag{16}$$

We then find, as in Chap. VI, that in Ampère's law in a magnetic medium, we must include the effects of all currents, both real currents and those arising from polarization, so that (15) is replaced by

$$\text{curl } \mathbf{B} = 4\pi(\mathbf{J} + \mathbf{J}') = 4\pi\mathbf{J} + 4\pi \text{ curl } \mathbf{M},$$
$$\text{curl } (\mathbf{B} - 4\pi\mathbf{M}) = 4\pi\mathbf{J}.$$

We then define a magnetic field **H,** by the equation

$$\mathbf{H} = \mathbf{B} - 4\pi\mathbf{M},$$

in terms of which Ampère's law becomes

$$\text{curl } \mathbf{H} = 4\pi\mathbf{J}. \tag{17}$$

We see that this definition is so set up that $\mathbf{B} = \mathbf{H}$ in a nonmagnetic medium. The relation between **B** and **H** may be expressed in the forms

$$\mathbf{B} = \mathbf{H} + 4\pi\mathbf{M} = \kappa_m\mathbf{H}, \qquad \kappa_m = 1 + 4\pi\chi_m, \quad \text{where } \mathbf{M} = \chi_m\mathbf{H}. \tag{18}$$

These relations are clearly analogous to the corresponding electrostatic relations (7) and (8).

Faraday's law, in unrationalized emu, takes the forms

$$\int \mathbf{E} \cdot \mathbf{ds} = -\frac{d}{dt} \int \mathbf{B} \cdot \mathbf{n} \, da, \qquad \text{curl } \mathbf{E} = -\frac{\partial \mathbf{B}}{\partial t}. \tag{19}$$

To see that these formulas, without additional constants, give the correct value, where **E** and **B** are expressed in emu, we may use the familiar derivation of Faraday's law from the energy principle, a derivation that we did not give in the body of the text. In this derivation, the flux through a circuit is changed by deforming the circuit, keeping the value of **B** constant in time, rather than by changing **B** with a constant circuit. The force exerted by the magnetic field on the element of length **ds** of wire carrying current i is $i(\mathbf{ds} \times \mathbf{B})$, and the work done by this field, if the wire is displaced by an amount **du,** is $i\int(\mathbf{ds} \times \mathbf{B}) \cdot \mathbf{du} = i\int\mathbf{B} \cdot (\mathbf{du} \times \mathbf{ds})$. The quantity $\mathbf{du} \times \mathbf{ds}$ is a vector equal in magnitude to the increment of area of the circuit, and pointing along the normal **n** to the surface spanning the circuit. Thus the work done, as computed above, may be rewritten $id(\int \mathbf{B} \cdot \mathbf{n} \, da)$. To do this external work, the current flowing in the wire must have lost a compensating amount of energy, which means that an emf $\int \mathbf{E} \cdot \mathbf{ds}$ must have existed, satisfying the relation

$$id(\int \mathbf{B} \cdot \mathbf{n} \, da) = -i \, dt \int \mathbf{E} \cdot \mathbf{ds},$$

from which Faraday's law follows at once. From Faraday's law, and the Biot-Savart law (11), we see at once that Eq. (2.2) of Chap. VII, for the coefficient of self- or mutual induction, is written in unrationalized emu in the form

$$L \text{ or } M = \int \int \frac{\mathbf{J} \cdot \mathbf{J}'}{ii'r} \, dv \, dv' \tag{20}$$

from which we see that the formulas of Chap. VII for inductances can be converted to emu by dividing by the factor $\mu_0/4\pi$.

We have now seen how to express the electrical quantities in unrationalized esu, and the magnetic and electrical quantities in unrationalized emu. When we combine these, to get Maxwell's equations, it is customary not to use either of these systems of units consistently, but to combine them into the Gaussian units. In these units, we use emu for **B** and **H**, but esu for all other quantities, including current and charge. The electrostatic equations will then be as in esu, but the equations involving magnetic quantities will be changed. We have the relation (10) between esu and emu of charge, current, potential, and correspondingly of electric field. If we let

$$c = 2.998 \times 10^{10} \text{ cm/sec}, \tag{21}$$

we see that **J**, expressed in emu, is equal to $1/c$ times **J** expressed in esu, while **E** expressed in emu is equal to c times **E** expressed in esu. Thus Maxwell's equations in Gaussian units become

$$\text{curl } \mathbf{E} = -\frac{1}{c}\frac{\partial \mathbf{B}}{\partial t}, \qquad \text{div } \mathbf{B} = 0$$

$$\text{curl } \mathbf{H} = \frac{1}{c}\frac{\partial \mathbf{D}}{\partial t} + 4\pi \frac{\mathbf{J}}{c}, \qquad \text{div } \mathbf{D} = 4\pi\rho, \tag{22}$$

where we have included the displacement-current term, whose correctness may be inferred from the fact that it satisfies the equation of continuity, $\text{div } \mathbf{J} + \partial\rho/\partial t = 0$. These equations are to be supplemented by the constitutive equations,

$$\mathbf{D} = \kappa_e\mathbf{E}, \qquad \mathbf{B} = \kappa_m\mathbf{H}, \tag{23}$$

where κ_e and κ_m are as given in (7), (8), and (18), and where in Gaussian units they are often denoted by ϵ and μ, respectively. Ohm's law is still stated in the form $\mathbf{J} = \sigma\mathbf{E}$, where the value of the conductivity σ is the same in Gaussian units that it is in esu.

From Maxwell's equations in Gaussian units, we can proceed to introduce the potentials, set up wave equations, and consider energy density and the Poynting vector. For the potentials we have

$$\mathbf{B} = \text{curl } \mathbf{A}, \qquad \mathbf{E} = -\text{grad } \varphi - \frac{1}{c}\frac{\partial\mathbf{A}}{\partial t}, \tag{24}$$

in which φ and $\mathbf{A}$ are subject to the relation

$$\text{div } \mathbf{A} + \frac{\kappa_e\kappa_m}{c}\frac{\partial\varphi}{\partial t} = 0. \tag{25}$$

Then we find that the potentials satisfy the following equations, if κ_e and κ_m are constants independent of position:

$$\nabla^2\varphi - \frac{\kappa_e\kappa_m}{c^2}\frac{\partial^2\varphi}{\partial t^2} = -\frac{4\pi\rho}{\kappa_e}$$
$$\nabla^2\mathbf{A} - \frac{\kappa_e\kappa_m}{c^2}\frac{\partial^2\mathbf{A}}{\partial t^2} = -4\pi\kappa_m\frac{\mathbf{J}}{c}. \tag{26}$$

Similar wave equations are satisfied by the components of electric and magnetic field. They show that, in a region containing no charge or current, waves will be propagated with a velocity $c/\sqrt{\kappa_e\kappa_m}$, verifying our earlier statement that c represents the velocity of light in empty space. In a conducting medium in which $\mathbf{J} = \sigma\mathbf{E}$, and in which the charge density is zero, we have instead of (25)

$$\text{div } \mathbf{A} + \frac{4\pi\kappa_m\sigma\varphi}{c} + \frac{\kappa_e\kappa_m}{c}\frac{\partial\varphi}{\partial t} = 0, \tag{27}$$

and instead of (26)

$$\nabla^2\varphi - \frac{4\pi\kappa_m\sigma}{c^2}\frac{\partial\varphi}{\partial t} - \frac{\kappa_e\kappa_m}{c^2}\frac{\partial^2\varphi}{\partial t^2} = 0, \tag{28}$$

with a similar equation for $\mathbf{A}$.

When we consider plane wave solutions of the wave equation, we find that in empty space $\mathbf{E}$ (in esu) is numerically equal to $\mathbf{H}$ (in emu). In a medium in which κ_e and κ_m are different from unity,

$\sqrt{\kappa_e}\,\mathbf{E}$ is numerically equal to $\sqrt{\kappa_m}\,\mathbf{H}$. In a conducting medium, propagating a wave with angular frequency ω, κ_e is replaced by

$$\kappa_e - \frac{4\pi j\sigma}{\omega},$$

so that $\sqrt{\kappa_e - 4\pi j\sigma/\omega}\,E$ is equal to $\sqrt{\kappa_m}\,H$, and the complex index of refraction, instead of being $\sqrt{\kappa_e\kappa_m}$, equals $\sqrt{\kappa_m(\kappa_e - 4\pi j\sigma/\omega)}$. From these formulas, the properties of reflection from conducting surfaces, skin depth, etc., can easily be found, but it should be remembered that in Gaussian units we must use σ in esu.

From Maxwell's equations we can prove Poynting's theorem, which proves to be

$$\operatorname{div}\left[\frac{c}{4\pi}\,(\mathbf{E}\times\mathbf{H})\right] + \frac{\partial}{\partial t}\left[\frac{1}{8\pi}(\kappa_e E^2 + \kappa_m H^2)\right] = -\mathbf{E}\cdot\mathbf{J}. \quad (29)$$

From this we infer that Poynting's vector, in unrationalized Gaussian units, is $(c/4\pi)(\mathbf{E}\times\mathbf{H})$, and the electric- and magnetic-energy densities are $\kappa_e E^2/8\pi$ and $\kappa_m H^2/8\pi$, respectively. We note from our relation between the magnitudes of E and H, mentioned above, that the magnetic- and electric-energy densities in a plane wave are equal to each other. Applying this result to a spherical solution of the wave equation, we find that the rate of radiation from a dipole of amplitude M, angular frequency ω, is

$$\int S\,da = \frac{\omega^4 M^2}{3c^3}, \quad (30)$$

analogous to Eq. (4.2), Chap. XII. Similarly the scattering cross section of a scattering electron, analogous to (5.3), Chap. XII, is

$$\sigma = \frac{8\pi}{3}\frac{\omega^4}{(\omega_0^2 - \omega^2)^2 + (\omega g)^2}\left(\frac{e^2}{mc^2}\right)^2. \quad (31)$$

We have now indicated the equivalent, in unrationalized Gaussian units, of many of the most important formulas that have been encountered in mks units in the body of the text. Next we consider the practical units. These are based on the emu, but differ from them by arbitrary powers of 10, chosen to make the practical units of convenient size. The two most fundamental units are the ampere and the volt. These are arbitrarily defined by the following equations:

$$1 \text{ ampere} = 10^{-1} \text{ emu of current}$$
$$1 \text{ volt} = 10^8 \text{ emu of potential}. \quad (32)$$

It follows from this that in the practical system the rate of working, or power, when 1 amp flows through a potential difference of 1 volt, is 10^7 ergs/sec, or 1 joule/sec, or 1 watt, so that power in the practical system is measured in watts. The coulomb is of course defined in terms of the ampere, and the unit of electric field is ordinarily the volt per centimeter. Using (10), we see that

$$1 \text{ esu of charge} = \frac{1}{2.998 \times 10^9} \text{ coulombs,}$$
$$1 \text{ esu of potential} = 299.8 \text{ volts.} \tag{33}$$

We see by dividing that

$$1 \text{ ohm} = 10^9 \text{ emu of resistance}$$
$$= \frac{1}{(2.998)^2 \times 10^{11}} \text{ esu of resistance.} \tag{34}$$

From Coulomb's law (1) we now see that the force between two charges q and q', expressed in coulombs, at a distance of separation r, measured in centimeters, is $F = (2.998)^2 \times 10^{18} qq'/r^2$ dynes or $(2.998)^2 \times 10^{13} qq'/r^2$ newtons. If r is measured in meters, the force is

$$F = (2.998)^2 \times 10^9 \frac{qq'}{r^2} \text{ newtons.} \tag{35}$$

This is identical with Eq. (2.2) of Chap. I, $F = qq'/4\pi\epsilon_0 r^2$, so that $\epsilon_0 = 1/[4\pi \times (2.998)^2 \times 10^9] = 8.85 \times 10^{-12}$, as in Eq. (2.1) of Chap. I. We note that (2.2) of Chap. I, which is the basis of our treatment of electrostatics in rationalized mks units, nevertheless uses as a unit the coulomb, which is an unrationalized unit, being based on the unrationalized emu. The factor 4π which appears in the definition of ϵ_0 compensates for this.

Similarly from the law (9) of force between two current elements, we see that the force between lengths ds_1 and ds_2 of wire carrying currents of i_1 and i_2, in amperes, at distance r, is

$$d\mathbf{F} = 10^{-2} i_1 i_2 \frac{[\mathbf{ds_1} \times (\mathbf{ds_2} \times \mathbf{r})]}{|\mathbf{r}|^3} \quad \text{dynes}$$
$$= 10^{-7} i_1 i_2 \frac{[\mathbf{ds_1} \times (\mathbf{ds_2} \times \mathbf{r})]}{|\mathbf{r}|^3} \quad \text{newtons.} \tag{36}$$

These formulas hold whether distances are in centimeters or in meters, since the dimensions of length cancel. Thus we verify Eq. (1.4) of Chap. V, in which we see that μ_0 must be defined as in Eq. (1.2), Chap. V. When we come to magnetic fields, we find that the unit in actual use is the gauss, which as we have seen is the emu of magnetic induction **B**. If then we have a current i, in amperes, flowing in a length **ds** of wire, in centimeters, in a magnetic induction **B**, in gausses, the force **dF**, in dynes, is

$$d\mathbf{F} = 10^{-1}i(\mathbf{ds} \times \mathbf{B}) \qquad \text{dynes.}$$

Similarly if **ds** is in meters, **dF** in newtons, we have

$$d\mathbf{F} = 10^{-4}i(\mathbf{ds} \times \mathbf{B}) \qquad \text{newtons.}$$

It is evident that to eliminate the numerical factor 10^{-4} in the mks system we must use a unit of **B** 10^4 times as large as the gauss. This unit is the weber per square meter.

The weber is a unit of flux, not of magnetic induction. It is a practical unit, based on Faraday's law: a time rate of change of flux of 1 weber/sec through a circuit induces an emf of 1 volt in that circuit. We recall, however, that a time rate of change of flux of 1 gauss/sec through an area of 1 sq cm induces an emf of 1 emu. Since 1 volt $= 10^8$ emu, this means that it requires a change of flux of 10^8 gausses per second flowing through an area of 1 sq cm, or of 10^4 gausses/sec flowing through 1 sq m, to induce 1 volt. In other words,

$$1 \text{ weber/sq cm} = 10^8 \text{ gausses,}$$
$$1 \text{ weber/sq m} = 10^4 \text{ gausses.} \qquad (37)$$

Thus we establish the relation between the weber and the gauss.

We have now discussed the practical units of electrical and magnetic quantities. At the same time, we have established the laws of force between charges and currents, and the other fundamental relations on which our treatment of the mks units in the text is based. We have not taken up the lucky accident by which the powers of 10 in terms of which the practical units of current and voltage are related to the emu, and the powers of 10 by which the meter and the newton are related to the centimeter and the dyne, are just such that they can be combined into a consistent set of units involving no powers of 10 in the final equations. This is in fact an accident, but when advantage is taken of it, we are led to the set of units taken up in the main discussion in the text. This set of units is still unfamiliar to many readers, who were brought up to use the Gaussian units. The writers believe firmly that the mks units are to be preferred to the Gaussian units, not only for practical calculations, but for all purposes. The equations expressed in terms of mks units are certainly no more complicated, and in many ways they are simpler, than those expressed in terms of Gaussian units. To that is added the great advantage that, whenever results are to be translated into practice, the units to be used in the mks system are those in practical use, whereas in the Gaussian system they are all unfamiliar, and a conversion table must be used. The only constants that need be remembered in the mks system are ϵ_0 and μ_0, and these have definite physical meanings: ϵ_0 measures the capacity of a condenser whose plates are each 1 m square, and 1 m apart; μ_0 measures the self-inductance, per meter length, of a transmission line consisting of two parallel-plane conductors, each 1 m broad, spaced 1 m apart. (Each of these definitions, of course, implies suitable conductors outside those in question, to prevent fringing of fields.)

APPENDIX III

FOURIER SERIES

Fourier's theorem may be stated as follows: Given an arbitrary function $\varphi(x)$. Then [unless $\varphi(x)$ contains an infinite number of discontinuities in a finite range, or similarly misbehaves itself], we can write

$$\varphi(x) = \frac{A_0}{2} + \sum_{n=1}^{\infty} \left(A_n \cos \frac{2\pi nx}{X} + B_n \sin \frac{2\pi nx}{X} \right) \tag{1}$$

where

$$A_n = \frac{2}{X} \int_{-X/2}^{X/2} \varphi(x) \cos \frac{2\pi nx}{X} \, dx,$$

$$B_n = \frac{2}{X} \int_{-X/2}^{X/2} \varphi(x) \sin \frac{2\pi nx}{X} \, dx. \tag{2}$$

This equation holds for values of x between $-X/2$ and $X/2$, but not in general outside this range. The series of sines and cosines is called "Fourier's series." There are two sides to the proof of Fourier's theorem. First, we may prove that, if a series of sines and cosines of this sort can represent the function, then it must have the coefficients we have given. This is simple, and we shall carry it through. But second, we could show that the series we so set up actually represents the function. That is, we should investigate the convergence of the series, show that it does converge, and that its sum is the function $\varphi(x)$. This second part we shall omit, merely stating the results of the discussion.

Let us suppose that $\varphi(x)$ is represented by a series as in (1), and ask what values the A's and B's must have if the equation is to be true. Multiply both sides of the equation by $\cos (2\pi mx/X)$, where m is an integer, and integrate from $-X/2$ to $X/2$. We then have

$$\int_{-X/2}^{X/2} \varphi(x) \cos \frac{2\pi mx}{X} \, dx = \int_{-X/2}^{X/2} \left(\frac{A_0}{2} \cos \frac{2\pi mx}{X} \right.$$
$$\left. + \sum_n A_n \cos \frac{2\pi nx}{X} \cos \frac{2\pi mx}{X} + \sum_n B_n \sin \frac{2\pi nx}{X} \cos \frac{2\pi mx}{X} \right) dx. \tag{3}$$

But we can easily show by direct integration that

$$\int_{-X/2}^{X/2} \cos \frac{2\pi nx}{X} \cos \frac{2\pi mx}{X} \, dx = 0$$

if n and m are integers, unless $n = m$, and that

$$\int_{-X/2}^{X/2} \sin \frac{2\pi nx}{X} \cos \frac{2\pi mx}{X} \, dx = 0$$

if n and m are integers. Thus all terms on the right of (3) are zero except one, for which $n = m$. The first term falls in with this rule, when we remember that $\cos 0 = 1$. This one term then gives us

$$A_n \int_{-X/2}^{X/2} \cos^2 \frac{2\pi nx}{X} \, dx = A_n \frac{X}{2},$$

as we can readily show. Hence

$$A_n = \frac{2}{X} \int_{-X/2}^{X/2} \varphi(x) \cos \frac{2\pi nx}{X} \, dx.$$

In a similar way, multiplying by $\sin (2\pi mx/X)$, we can prove the formula for B_n.

We have thus shown that, if a function $\varphi(x)$ is to be represented by a series (1), the coefficients must be given by (2). We shall next make a few remarks about the other part of the problem, the question whether the series so defined really converges to represent the function $\varphi(x)$. In the first place, the series cannot in general represent the function, except in the region between $-X/2$ and $X/2$. For the series is periodic, repeating itself in every period, whereas the function in general is not. Only periodic functions of this period can be represented in all their range by Fourier series. If we try to represent a nonperiodic function, the representation will be correct within the range from $-X/2$ to $X/2$, but the same thing will automatically repeat itself outside the range. Incidentally, we can easily change the range in which the series represents the function. If we merely change the range of integration so as to be from x_0 to $x_0 + X$, where x_0 is arbitrary, the series will represent the function within this range. The case we have used corresponds to $x_0 = -X/2$; another choice frequently made is $x_0 = 0$.

We can also change the value of X, and thereby change the length of the range in which the series is correct. To represent a function through a large range of x, we may use a large value of X. In fact, as X becomes infinite, the quantities $2\pi nx/X$ for successive n's become arbitrarily close together, and the summations involved in (1) and (2) may be replaced by integrations. Thus, if we let $2\pi nx/X$ equal ω, the interval $d\omega$ between successive values of this quantity will be $d\omega = 2\pi/X$. We may then replace (1) by the integration

$$\varphi(x) = \frac{X}{2\pi} \int_0^\infty (A_n \cos \omega x + B_n \sin \omega x) \, d\omega,$$

where

$$A_n = \frac{2}{X} \int_{-\infty}^\infty \varphi(\xi) \cos \omega \xi \, d\xi,$$

$$B_n = \frac{2}{X} \int_{-\infty}^\infty \varphi(\xi) \sin \omega \xi \, d\xi.$$

Here we have left out the term in A_0, which becomes negligible in the limit. Combining, we finally have

$$\varphi(x) = \frac{1}{\pi} \int_0^\infty d\omega \int_{-\infty}^\infty \varphi(\xi) \cos \omega(\xi - x) \, d\xi. \tag{4}$$

This theorem expresses Fourier's integral theorem, which as we see is merely the limiting form of Fourier's series for infinite value of the period X.

There is an alternative form of Fourier's theorem (1) and (2), expressed in terms of exponential rather than trigonometric functions, which is simpler to write, and for many purposes is more convenient. To derive it, we express the sines and cosines in (1) and (2) in complex exponential form. Grouping together the terms in $e^{j2\pi nx/X}$, and the terms in $e^{-j2\pi nx/X}$, we find without trouble that an equivalent statement of the theorem is

$$\varphi(x) = \sum_{n=-\infty}^\infty C_n e^{j2\pi nx/X} \tag{5}$$

where

$$C_n = \frac{1}{X} \int_{-X/2}^{X/2} \varphi(x) e^{-j2\pi nx/X} \, dx. \tag{6}$$

The terms for equal positive and negative values of n combine to give the sines and cosines correctly, and the term for $n = 0$ gives the term in A_0 in (1). Though the C_n's, from their definition, are complex, it is easily shown that, if $\varphi(x)$ is real, C_n and C_{-n} are complex conjugates, and the series itself is real. We can easily write an expression of Fourier's integral theorem, equivalent to (4), but in exponential language, similar to (5) and (6).

Although the range within which a Fourier series converges to the value of the function it is supposed to represent is limited, as we have seen, to the value X, there is a compensation, in that within this range a Fourier series can be used to represent much worse curves than a power series. Thus a Fourier series can still converge, even though the function has a finite number of finite discontinuities. It can consist, for example, of one function in one part of the region, another in another (in this case, to carry out the integrations, we must break up the integral into separate integrals over these parts, and add them). The less serious the discontinuities, however, the better the convergence. Thus, if the function itself has discontinuities, the coefficients will fall off as $1/n$; whereas, if only the first derivative has discontinuities, the coefficients will fall off as $1/n^2$, etc. Differentiating a function makes the convergence of a series worse, as we can see, for example, if a function is continuous but its first derivative is discontinuous. Then the coefficients fall off as $1/n^2$, but if we differentiate, the coefficients of the resulting series will fall off as $1/n$. There is an interesting point connected with the series for a discontinuous function. If the function jumps from one value u_1 to another u_2 at a given value of x, then the series at this point converges to the mean value, $(u_1 + u_2)/2$.

Problems

1. Expand in Fourier series the function that is equal to $-x$ for x between $-X/2$ and 0, and equal to x for x between 0 and $X/2$.

2. Expand in Fourier series the function that is equal to -1 for x between $-X/2$ and 0, and equal to 1 for x between 0 and $X/2$. See if this series can be found by differentiating the series of Prob. 1 term by term. Consider the convergence of these two series, with reference to their continuity. What happens if we try to differentiate again term by term?

3. Expand in Fourier series the function that is equal to x^2 for x between $-X/2$ and $X/2$. Compute the sum of the first four terms of this series, and see how good an approximation to the function you have.

4. Expand in Fourier series the function that is equal to zero except for x between $-\xi/2$ and $\xi/2$, where $\xi \ll X$, while in this region it equals unity. Discuss the behavior of this function in the limit as ξ becomes zero.

APPENDIX IV

VECTOR OPERATIONS IN CURVILINEAR COORDINATES

Let us assume three orthogonal coordinates q_1, q_2, q_3, so that the three sets of coordinate surfaces, $q_1 =$ constant, $q_2 =$ constant, $q_3 =$ constant, intersect at right angles, though in general the surfaces will be curved. Now let us move a distance ds_1 normal to a surface $q_1 =$ constant. By doing so, q_2 and q_3 do not change, but we reach another surface on which q_1 has increased by dq_1, which in general is different from ds_1. Thus, with polar coordinates, if the displacement is along the radius, so that r is changing, $ds = dr$; but if it is along a tangent to a circle, so that θ is changing, $ds = r\,d\theta$. In general, we have

$$ds_1 = h_1\,dq_1, \qquad ds_2 = h_2\,dq_2, \qquad ds_3 = h_3\,dq_3, \qquad (1)$$

where in polar coordinates the h connected with r is unity, but that connected with θ is r. The first step in setting up vector operations in any set of coordinates is to derive these h's, which can be done by elementary geometrical methods. Thus in cylindrical coordinates, where the coordinates are r, θ, z, we have $ds_1 = dr$, $ds_2 = r\,d\theta$, $ds_3 = dz$, so that $h_1 = 1$, $h_2 = r$, $h_3 = 1$. In spherical polar coordinates, r, θ, φ, we have $ds_1 = dr$, $ds_2 = r\,d\theta$, $ds_3 = r\sin\theta\,d\varphi$, so that $h_1 = 1$, $h_2 = r$, $h_3 = r\sin\theta$.

Gradient.—The component of the gradient of a scalar S in any direction is its rate of change in that direction. Thus the component in the direction 1 (normal to the surface $q_1 =$ constant) is

$$\frac{dS}{ds_1} = \left(\frac{1}{h_1}\right)\left(\frac{\partial S}{\partial q_1}\right),$$

with similar formulas for the other components. Thus in cylindrical coordinates we have

$$\operatorname{grad}_r S = \frac{\partial S}{\partial r}, \qquad \operatorname{grad}_\theta S = \frac{1}{r}\frac{\partial S}{\partial \theta}, \qquad \operatorname{grad}_z S = \frac{\partial S}{\partial z}, \qquad (2)$$

and in spherical coordinates we have

$$\operatorname{grad}_r S = \frac{\partial S}{\partial r}, \qquad \operatorname{grad}_\theta S = \frac{1}{r}\frac{\partial S}{\partial \theta}, \qquad \operatorname{grad}_\varphi S = \frac{1}{r\sin\theta}\frac{\partial S}{\partial \varphi}. \qquad (3)$$

Divergence.—Let us apply the divergence theorem to a small volume element $dV = ds_1\, ds_2\, ds_3$, bounded by coordinate surfaces at q_1, $q_1 + dq_1$, etc. If we have a vector **A**, with components A_1, A_2, A_3 along the three curvilinear axes, the flux into the volume over the face at q_1, whose area is $ds_2\, ds_3$, is $(A_1\, ds_2\, ds_3)_{q_1}$, and the corresponding flux out over the opposite face is $(A_1\, ds_2\, ds_3)_{q_1+dq_1}$, where we note that the area $ds_2\, ds_3$ changes with q_1 as well as the flux density A_1. Thus the flux out over these two faces is

$$\frac{\partial}{\partial q_1}(A_1\, ds_2\, ds_3)\, dq_1 = \frac{\partial}{\partial q_1}(A_1 h_2 h_3)\, dq_1\, dq_2\, dq_3 = \frac{1}{h_1 h_2 h_3}\frac{\partial}{\partial q_1}(A_1 h_2 h_3)\, dv.$$

Proceeding similarly with the other pairs of faces, and setting the whole outward flux equal to div **A** dv, we have

$$\text{div } \mathbf{A} = \frac{1}{h_1 h_2 h_3}\left[\frac{\partial}{\partial q_1}(A_1 h_2 h_3) + \frac{\partial}{\partial q_2}(A_2 h_3 h_1) + \frac{\partial}{\partial q_3}(A_3 h_1 h_2)\right]. \quad (4)$$

Thus in cylindrical coordinates we have

$$\text{div } \mathbf{A} = \frac{1}{r}\frac{\partial}{\partial r}(rA_r) + \frac{1}{r}\frac{\partial A_\theta}{\partial \theta} + \frac{\partial A_z}{\partial z} \quad (5)$$

and in spherical coordinates

$$\text{div } \mathbf{A} = \frac{1}{r^2}\frac{\partial}{\partial r}(r^2 A_r) + \frac{1}{r\sin\theta}\frac{\partial}{\partial \theta}(\sin\theta\, A_\theta) + \frac{1}{r\sin\theta}\frac{\partial A_\varphi}{\partial \varphi}. \quad (6)$$

Laplacian.—Writing the Laplacian of a scalar S as div grad S, and placing $A_1 = \text{grad}_1\, S$, etc., in the expression for div A, we have

$$\nabla^2 S = \frac{1}{h_1 h_2 h_3}\left[\frac{\partial}{\partial q_1}\left(\frac{h_2 h_3}{h_1}\frac{\partial S}{\partial q_1}\right) + \frac{\partial}{\partial q_2}\left(\frac{h_3 h_1}{h_2}\frac{\partial S}{\partial q_2}\right) + \frac{\partial}{\partial q_3}\left(\frac{h_1 h_2}{h_3}\frac{\partial S}{\partial q_3}\right)\right]. \quad (7)$$

Thus in cylindrical coordinates we have

$$\nabla^2 S = \frac{1}{r}\frac{\partial}{\partial r}\left(r\frac{\partial S}{\partial r}\right) + \frac{1}{r^2}\frac{\partial^2 S}{\partial \theta^2} + \frac{\partial^2 S}{\partial z^2} \quad (8)$$

and in spherical coordinates

$$\nabla^2 S = \frac{1}{r^2}\frac{\partial}{\partial r}\left(r^2\frac{\partial S}{\partial r}\right) + \frac{1}{r^2\sin\theta}\frac{\partial}{\partial \theta}\left(\sin\theta\frac{\partial S}{\partial \theta}\right) + \frac{1}{r^2\sin^2\theta}\frac{\partial^2 S}{\partial \varphi^2}. \quad (9)$$

Curl.—We apply Stokes's theorem to an approximately rectangular area bounded by q_1, $q_1 + dq_1$, q_2, $q_2 + dq_2$. The line integral of a vector **A** about the circuit is

$$A_1(q_1,q_2)\ ds_1 + A_2(q_1 + dq_1,q_2)\ ds_2 - A_1(q_1,q_2 + dq_2)\ ds_1 - A_2(q_1,q_2)\ ds_2.$$

This is approximately equal to

$$\left[\frac{\partial}{\partial q_1}(h_2 A_2) - \frac{\partial}{\partial q_2}(h_1 A_1)\right] dq_1\ dq_2. \tag{10}$$

Since this must be $\mathrm{curl}_3\ \mathbf{A}\ ds_1\ ds_2$, we have

$$\mathrm{curl}_3\ \mathbf{A} = \frac{1}{h_1 h_2}\left[\frac{\partial}{\partial q_1}(h_2 A_2) - \frac{\partial}{\partial q_2}(h_1 A_1)\right], \tag{11}$$

with similar expressions for the other components. Thus in cylindrical coordinates we have

$$\mathrm{curl}_r\ \mathbf{A} = \frac{1}{r}\frac{\partial A_z}{\partial \theta} - \frac{\partial A_\theta}{\partial z}$$

$$\mathrm{curl}_\theta\ \mathbf{A} = \frac{\partial A_r}{\partial z} - \frac{\partial A_z}{\partial r}$$

$$\mathrm{curl}_z\ \mathbf{A} = \frac{1}{r}\frac{\partial}{\partial r}(r A_\theta) - \frac{1}{r}\frac{\partial A_r}{\partial \theta} \tag{12}$$

and in spherical coordinates

$$\mathrm{curl}_r\ \mathbf{A} = \frac{1}{r \sin\theta}\frac{\partial}{\partial \theta}(\sin\theta\ A_\varphi) - \frac{1}{r \sin\theta}\frac{\partial A_\theta}{\partial \varphi}$$

$$\mathrm{curl}_\theta\ \mathbf{A} = \frac{1}{r \sin\theta}\frac{\partial A_r}{\partial \varphi} - \frac{1}{r}\frac{\partial}{\partial r}(r A_\varphi)$$

$$\mathrm{curl}_\varphi\ \mathbf{A} = \frac{1}{r}\frac{\partial}{\partial r}(r A_\theta) - \frac{1}{r}\frac{\partial A_r}{\partial \theta}. \tag{13}$$

APPENDIX V

SPHERICAL HARMONICS

In Chap. III, we considered the solution of Laplace's equation in spherical coordinates, and showed that the solution may be written as a product of a function of r, a function of θ, and a function of φ. The function of φ is $\sin m\varphi$ or $\cos m\varphi$, and the function Θ of θ satisfies the differential equation

$$\frac{1}{\sin \theta} \frac{d}{d\theta}\left(\sin \theta \frac{d\Theta}{d\theta}\right) + \left[l(l+1) - \frac{m^2}{\sin^2 \theta}\right] \Theta = 0. \tag{1}$$

We showed that the function Θ could be written in the form

$$\Theta = \sin^m \theta (A_0 + A_2 \cos^2 \theta + A_4 \cos^4 \theta + \cdots)$$

or

$$= \sin^m \theta (A_1 + A_3 \cos^3 \theta + A_5 \cos^5 \theta + \cdots), \tag{2}$$

where

$$\frac{A_n}{A_{n+2}} = -\frac{(n+1)(n+2)}{l(l+1) - (m+n)(m+n+1)}. \tag{3}$$

We found further that if l is an integer one or the other of the series in (2) breaks off to form a polynomial. On the other hand, we stated without proof that, if the series does not break off, it leads to a function that becomes infinite when $\cos \theta = \pm 1$, so that such a function cannot be used for expanding solutions of Laplace's equation that remain finite at $\cos \theta = \pm 1$. Thus it is only the polynomial solutions that are usually of use.

These polynomials, expressed in a certain form, with suitably chosen values of the coefficients, are the associated Legendre polynomials. They are ordinarily expressed, not as in (2) in a series of increasing powers of $\cos \theta$, but in a series of descending powers, starting with the term of highest power. In this form, they are defined as

$$P_l^m(\cos \theta) = \frac{\sin^m \theta (2l)!}{2^l l! (l-m)!}\left[(\cos \theta)^{l-m} - \frac{(l-m)(l-m-1)}{2(2l-1)}(\cos \theta)^{l-m-2}\right.$$
$$+ \frac{(l-m)(l-m-1)(l-m-2)(l-m-3)}{2 \cdot 4(2l-1)(2l-3)}(\cos \theta)^{l-m-4} - \cdots$$
$$+ \frac{(-1)^p(l-m)(l-m-1)\cdots(l-m-2p+1)}{2 \cdot 4 \cdots (2p)(2l-1)(2l-3)\cdots(2l-2p+1)}(\cos \theta)^{l-m-2p}$$
$$\left. + \cdots \right]. \tag{4}$$

To show the identity of this expression with (2), we need merely show that the ratio of successive coefficients is as given in (3). To do this, we may identify the term in $(\cos \theta)^{l-m-2p}$ in (4) with the term in $(\cos \theta)^n$ in (2), and the term in $(\cos \theta)^{l-m-2p+2}$ in (4) with that in $(\cos \theta)^{n+2}$ in (2). Doing this, taking the ratio of coefficients from (4), we find easily that it is the same as given by (3), so that (4) forms a legitimate way of writing the solution (2). The constant factor multiplying the series in (4) is chosen to simplify certain formulas.

For certain purposes it is more convenient to let $\cos \theta = x$, and to express the associated Legendre polynomials in terms of x. The equation (1), when expressed in terms of x, is

$$(1 - x^2) \frac{d^2\Theta}{dx^2} - 2x \frac{d\Theta}{dx} + \left[l(l + 1) - \frac{m^2}{1 - x^2} \right] \Theta = 0. \quad (5)$$

The function (4) can be written in a form that is sometimes useful, and that can be justified directly from the differential equation (5). This is

$$P_l^m(x) = \frac{1}{2^l l!} (1 - x^2)^{m/2} \frac{d^{l+m}}{dx^{l+m}} (x^2 - 1)^l. \quad (6)$$

From the form (6), we can arrive at the expression (4) by expanding $(x^2 - 1)^l$ in binomial expansion, and carrying out the differentiation term by term. The Legendre functions are the special case of (4) or (6) for which $m = 0$. They are often denoted $P_l(\cos \theta)$ or $P_l(x)$.

The associated Legendre polynomials have many important properties, but for our present purposes the most useful ones are their orthogonality and normalization relations. It can be proved that

$$\int_{-1}^{1} P_l^m(x) P_n^m(x) \, dx = 0 \qquad \text{if } l \neq n$$
$$= \frac{2}{2l + 1} \frac{(l + m)!}{(l - m)!} \text{if } l = n. \quad (7)$$

This relation can be used in the following way to satisfy boundary conditions over the surface of a sphere, in solutions of Laplace's equation: As we see from Eq. (3.3), Chap. III, such a solution can be expressed as a function of the angle by

$$\psi = \sum_{lm} P_l^m (\cos \theta)(A_{lm} \sin m\varphi + B_{lm} \cos m\varphi). \quad (8)$$

In this summation, l is to go from 0 to infinity, m from 0 to l (for from our definitions we see that m can never be greater than l). Let us suppose that the sum (8) is to be a specified function of θ and φ, which we may write $\psi_0(\cos\theta,\varphi)$. Let us then multiply both sides of (8) by a particular function, say $P_p^n(\cos\theta)\sin n\varphi$, and integrate over all solid angles (which amounts to integrating over the surface of the sphere). The element of solid angle is $\sin\theta\,d\theta\,d\varphi$. If we let $x = \cos\theta$, we have $dx = -\sin\theta\,d\theta$. Thus we have

$$\int_{-1}^{1} dx \int_{0}^{2\pi} d\varphi\, P_p^n(x)\sin n\varphi\, \psi_0(x,\varphi) = \sum_{lm} \int_{-1}^{1} P_l^m(x)P_p^n(x)\,dx$$
$$\int_{0}^{2\pi} \sin n\varphi(A_{lm}\sin m\varphi + B_{lm}\cos m\varphi)\,d\varphi.$$

Because of the orthogonality of the sines and cosines, the integral over φ on the right is zero unless $n = m$, in which case it is πA_{lm}. Setting $n = m$, the integral over x on the right is zero, because of (7), unless $l = p$, in which case its value is given by (7). Thus we have

$$\frac{2\pi}{2l + 1}\frac{(l + m)!}{(l - m)!}A_{lm} = \int_{-1}^{1} dx \int_{0}^{2\pi} P_p^n(x)\sin n\varphi\,\psi_0(x,\varphi)\,d\varphi. \quad (9)$$

In a similar way we can find a formula for B_{lm}.

APPENDIX VI

MULTIPOLES

The properties of multipoles can be conveniently considered either in rectangular or in spherical coordinates. We shall give both types of discussion, later indicating briefly the relation between them. First we define the dipole and quadrupole moments of a distribution of charges. Suppose we have a collection of point charges, all located near the origin. Let the ith charge be e_i, and let it be located at a point with coordinates ξ_i, η_i, ζ_i. Then the dipole moment is a vector, which we may denote by $\mathbf{p}^{(1)}$, with components

$$p_x^{(1)} = \sum_i e_i \xi_i, \qquad p_y^{(1)} = \sum_i e_i \eta_i, \qquad p_z^{(1)} = \sum_i e_i \zeta_i. \tag{1}$$

It is easy to see that this is consistent with the definition of dipole moment given in the text, by considering for instance a dipole consisting of a charge $+e$ at point d along the x axis, and charge $-e$ at the origin; its dipole moment is then a vector along the x axis, of magnitude ed. Similarly the quadrupole moment is a tensor. (The reader unfamiliar with tensor notation will find a discussion in Appendix V of *Mechanics*, by J. C. Slater and N. H. Frank, McGraw-Hill Book Company, Inc., New York, 1947.) We may denote it by $p^{(2)}$, and its components are

$$p_{xx}^{(2)} = \sum_i e_i \xi_i^2, \qquad p_{yy}^{(2)} = \sum_i e_i \eta_i^2, \qquad p_{zz}^{(2)} = \sum_i e_i \zeta_i^2,$$

$$p_{xy}^{(2)} = \sum_i e_i \xi_i \eta_i, \qquad p_{yz}^{(2)} = \sum_i e_i \eta_i \zeta_i, \qquad p_{zx}^{(2)} = \sum_i e_i \zeta_i \xi_i. \tag{2}$$

In a similar way one can define an octopole moment $p^{(3)}$, whose components are products of three displacements, and which therefore has three subscripts, etc.

We encounter the various moments of the charge distribution both in finding the field of the distribution at distant points, and in finding its potential energy in an arbitrary potential field. The electrostatic potential ψ at point x, y, z, arising from the charge distribution, by

Coulomb's law, is

$$\psi = \frac{1}{4\pi\epsilon_0} \sum_i \frac{e_i}{\rho_i}, \qquad (3)$$

where $\rho_i = \sqrt{(x - \xi_i)^2 + (y - \eta_i)^2 + (z - \zeta_i)^2}$. Assuming that x, y, z are larger than ξ_i, η_i, ζ_i, we may expand $1/\rho_i$, in (3), in Taylor's series in ξ_i, η_i, ζ_i. In doing this, we must take the derivatives of ρ_i with respect to ξ_i, η_i, ζ_i, and then set these quantities equal to zero. We note, however, that the derivative of ρ_i with respect to ξ_i is the negative of its derivative with respect to x, etc. We may then replace the derivatives by the negative derivatives of $r = \sqrt{x^2 + y^2 + z^2}$ with respect to x, y, z. Proceeding in this way, we find that

$$\psi = \frac{1}{4\pi\epsilon_0} \left\{ \frac{\sum_i e_i}{r} - \left[p_x^{(1)} \frac{\partial(1/r)}{\partial x} + p_y^{(1)} \frac{\partial(1/r)}{\partial y} + p_z^{(1)} \frac{\partial(1/r)}{\partial z} \right] \right.$$
$$+ \frac{1}{2} p_{xx}^{(2)} \frac{\partial^2(1/r)}{\partial x^2} + \frac{1}{2} p_{yy}^{(2)} \frac{\partial^2(1/r)}{\partial y^2} + \frac{1}{2} p_{zz}^{(2)} \frac{\partial^2(1/r)}{\partial z^2}$$
$$\left. + p_{xy}^{(2)} \frac{\partial^2(1/r)}{\partial x \, \partial y} + p_{yz}^{(2)} \frac{\partial^2(1/r)}{\partial y \, \partial z} + p_{zx}^{(2)} \frac{\partial^2(1/r)}{\partial z \, \partial x} + \cdots \right\}. \quad (4)$$

Here the first term represents the potential of the total charge, as if it were concentrated at the origin, the next three terms represent the potential of the dipole moment, the next of the quadrupole moment, etc.

The expression (4) is easier to interpret if we note the form of the various derivatives. We have

$$\frac{\partial(1/r)}{\partial x} = -\frac{x}{r^3}, \qquad \frac{\partial^2(1/r)}{\partial x^2} = -\frac{1}{r^3} + \frac{3x^2}{r^5}, \qquad \frac{\partial^2(1/r)}{\partial x \, \partial y} = \frac{3xy}{r^5},$$

with similar formulas for the other derivatives. Thus we see that the potential arising from the dipole moment is

$$\psi = \frac{1}{4\pi\epsilon_0} \frac{[p_x^{(1)}x + p_y^{(1)}y + p_z^{(1)}z]}{r^3}. \qquad (5)$$

In the numerator we have the scalar product of the dipole moment and the radius vector; since this is the magnitude of the dipole moment, times r, times the cosine of the angle between, we verify Eq. (5.1) of Chap. III for the potential of a dipole. We note further that the potential arising from a dipole is proportional to $1/r^2$ times a function of the angle, and by examination of (4) and (5) we find similarly that the potential arising from a quadrupole is proportional to $1/r^3$ times a function of angle, with correspondingly increasing exponents of the inverse powers for the higher multipoles.

The type of expansion we have used can be employed to find the potential energy of a charge distribution in an external field. Let there be a potential function $\psi(x,y,z)$ arising from charges external to the distribution we are considering. Then the potential energy of the distribution in this external field is

$$V = \sum_i e_i \psi(\xi_i, \eta_i, \zeta_i). \tag{6}$$

Expanding $\psi(\xi_i, \eta_i, \zeta_i)$ by Taylor's theorem, this becomes

$$V = \left(\sum_i e_i \right) \psi(0) + p_x^{(1)} \frac{\partial \psi}{\partial x} + p_y^{(1)} \frac{\partial \psi}{\partial y} + p_z^{(1)} \frac{\partial \psi}{\partial z}$$
$$+ \frac{1}{2} p_{xx}^{(2)} \frac{\partial^2 \psi}{\partial x^2} + \frac{1}{2} p_{yy}^{(2)} \frac{\partial^2 \psi}{\partial y^2} + \frac{1}{2} p_{zz}^{(2)} \frac{\partial^2 \psi}{\partial z^2}$$
$$+ p_{xy}^{(2)} \frac{\partial^2 \psi}{\partial x \, \partial y} + p_{yz}^{(2)} \frac{\partial^2 \psi}{\partial y \, \partial z} + p_{zx}^{(2)} \frac{\partial^2 \psi}{\partial z \, \partial x} + \cdots . \tag{7}$$

We note that the dipole terms in the potential energy form the scalar product of the dipole moment and the gradient of the potential, or the negative of the scalar product of dipole moment and electric field. By investigating the change of V, in (7), when the charge distribution is given a translation or a rotation, we can find the forces and torques acting on the charge distribution. We find, of course, that there is no net force acting on a dipole in a uniform field; the force arises only when the field strength varies with position. There is, however, clearly a torque acting on a dipole in a uniform field, tending to force the dipole into parallelism with the field.

We shall now express the formulation of the potential arising from a charge distribution in spherical polar coordinates. Let the coordinates of the charge e_i be r_i, θ_i, φ_i, and the coordinates of the point where we wish to find the potential r, θ, φ. Then, writing x, y, z, ξ_i, η_i, ζ_i, in terms of spherical coordinates, we find that the distance between the point r, θ, φ, and the point r_i, θ_i, φ_i where the ith charge is located is

$$\rho_i = \sqrt{r^2 - 2rr_i[\cos\theta\cos\theta_i + \sin\theta\sin\theta_i\cos(\varphi - \varphi_i)] + r_i^2}. \tag{8}$$

We then substitute in (3) to get the potential. In deriving this, we shall make use of a very important theorem relating to spherical harmonics, whose proof we shall not give. This is the following:

$$\frac{1}{\rho_i} = \sum_{l=0}^{\infty} \frac{r_i^l}{r^{l+1}} \left[P_l(\cos \theta) P_l(\cos \theta_i) \right.$$
$$\left. + 2 \sum_{m=1}^{l} \frac{(l-m)!}{(l+m)!} P_l^m(\cos \theta) P_l^m(\cos \theta_i) \cos m(\varphi - \varphi_i) \right] \quad (9)$$

if $r > r_i$. This expansion suggests setting up the following quantities, which serve as the various components of multipole moments in spherical coordinates:

$$p_0^{(l)} = \sum_i e_i r_i^l P_l(\cos \theta_i)$$

$$p_{m,1}^{(l)} = 2 \sum_i \frac{(l-m)!}{(l+m)!} e_i r_i^l P_l^m(\cos \theta_i) \cos m\varphi_i$$

$$p_{m,2}^{(l)} = 2 \sum_i \frac{(l-m)!}{(l+m)!} e_i r_i^l P_l^m(\cos \theta_i) \sin m\varphi_i. \quad (10)$$

We shall point out the nature of these quantities later, and shall show that those for $l = 1$ describe the dipole moment, those for $l = 2$ the quadrupole moment, etc. In terms of them, the potential is then

$$\psi = \frac{1}{4\pi\epsilon_0} \sum_l \frac{1}{r^{l+1}} \left\{ p_0^{(l)} P_l(\cos \theta) \right.$$
$$\left. + \sum_{m=1}^{l} [p_{m,1}^{(l)} P_l^m(\cos \theta) \cos m\varphi + p_{m,2}^{(l)} P_l^m(\cos \theta) \sin m\varphi] \right\}. \quad (11)$$

Let us now examine the expression (11), to find its significance. The term in $l = 0$ is very simple. Remembering, as follows from Appendix V, that $P_0(\cos \theta) = 1$, it is

$$\frac{1}{4\pi\epsilon_0} \frac{\sum_i e_i}{r}$$

expressing the potential of the total charge, as in the first term of (4). For the term in $l = 1$, we have $P_1(\cos \theta) = \cos \theta$, $P_1^1(\cos \theta) = \sin \theta$.

Then

$$p_0^{(1)} = \sum_i e_i r_i \cos \theta_i = \sum_i e_i \zeta_i$$

$$p_{1,1}^{(1)} = \sum_i e_i r_i \sin \theta_i \cos \varphi_i = \sum_i e_i \xi_i$$

$$p_{1,2}^{(1)} = \sum_i e_i r_i \sin \theta_i \sin \varphi_i = \sum_i e_i \eta_i. \tag{12}$$

Thus these three quantities are the components of the dipole moment, as we have defined them previously. Substituting in (11), we find that the term for $l = 1$ reduces to exactly the expression (5), when we remember that $x = r \sin \theta \cos \varphi$, $y = r \sin \theta \sin \varphi$, $z = r \cos \theta$. In a similar way we can identify the terms in $l = 2$ in (11) with the quadrupole terms in the rectangular case, though this identification is not so simple as with the dipole, and involves a paradox, in that there are only five components of the form given in (10), whereas we found six components of the quadrupole moment in (2). The paradox becomes resolved when we write out the potential in detail in both coordinate systems. Proceeding in a similar way, we see that the term of (11) corresponding to each value of l expresses the potential arising from a particular multipole.

APPENDIX VII

BESSEL'S FUNCTIONS

Bessel's equation is

$$\frac{1}{z} \frac{d}{dz} \left(z \frac{dZ}{dz} \right) + \left(1 - \frac{m^2}{z^2} \right) Z = 0, \tag{1}$$

and its solutions are $Z = J_m(z)$, $Z = N_m(z)$. By expanding Z in power series, we may show that

$$J_m(z) = \frac{1}{m!} \left(\frac{z}{2} \right)^m - \frac{1}{(m+1)!} \left(\frac{z}{2} \right)^{m+2} + \frac{1}{2!(m+2)!} \left(\frac{z}{2} \right)^{m+4} \cdots \tag{2}$$

where the coefficient of the first term is chosen according to an arbitrary convention. There is no similarly simple expansion for the Neumann function $N_m(z)$; it requires not only terms in positive and negative powers of z, but also logarithmic terms. For $N_0(z)$, the leading term for small values of z is the logarithmic term:

$$\lim_{z \to 0} N_0(z) = \frac{2}{\pi} (\ln z - 0.11593).$$

For the higher values of m, the term in inverse powers of z is the leading term for small z:

$$\lim_{z \to 0} N_m(z) = - \frac{(m-1)!}{\pi} \left(\frac{2}{z} \right)^m, \qquad m > 0. \tag{3}$$

For values of z large compared with m, the Bessel and Neumann functions have asymptotic expressions as follows:

$$\lim_{z \to \infty} J_m(z) = \sqrt{\frac{2}{\pi z}} \cos \left(z - \frac{2m+1}{4} \pi \right),$$
$$\lim_{z \to \infty} N_m(z) = \sqrt{\frac{2}{\pi z}} \sin \left(z - \frac{2m+1}{4} \pi \right). \tag{4}$$

An important relation connecting the Bessel and Neumann functions is

$$N_{m-1}(z) J_m(z) - N_m(z) J_{m-1}(z) = \frac{2}{\pi z}.$$

An important relation gives $J_m(z)$ as an integral:

$$J_m(z) = \frac{1}{2\pi j^m} \int_0^{2\pi} e^{jz \cos w} \cos (mw) \, dw,$$

where $j = \sqrt{-1}$.

In addition to these relations, there are a group of important relations holding equally well for either the Bessel or Neumann functions, and which we shall therefore state in terms of $Z_m(z)$, meaning by this either $J_m(z)$ or $N_m(z)$. For the Bessel functions, most of these relations can be proved without difficulty from the series representation. The relations are more general, however, in that they apply to the Neumann functions as well, and can be proved directly from the properties of the differential equation. They are the following:

$$Z_{m-1}(z) + Z_{m+1}(z) = \frac{2m}{z} Z_m(z)$$

$$Z_{-m}(z) = (-1)^m Z_m(z), \qquad \text{if } m \text{ is integral}$$

$$\frac{d}{dz} Z_m(z) = \frac{1}{2} [Z_{m-1}(z) - Z_{m+1}(z)]$$

$$\frac{d}{dz} [z^m Z_m(z)] = z^m Z_{m-1}(z)$$

$$\frac{d}{dz} [z^{-m} Z_m(z)] = -z^{-m} Z_{m+1}(z)$$

$$\int Z_1(z) \, dz = -Z_0(z)$$

$$\int z Z_0(z) \, dz = z Z_1(z)$$

$$\int Z_0^2(z) z \, dz = \frac{z^2}{2} [Z_0^2(z) + Z_1^2(z)]$$

$$\int Z_m^2(z) z \, dz = \frac{z^2}{2} [Z_m^2(z) - Z_{m-1}(z) Z_{m+1}(z)].$$

In Chap. XII, Sec. 2, we introduced spherical Bessel and Neumann functions, defined by the equations (2.1) of that chapter. We showed in Eq. (2.2) of that chapter that they can be expressed in terms of algebraic and trigonometric functions; we found their asymptotic behavior in (2.3), and their behavior for small z in (2.4), all of that chapter. These properties can be easily derived from the relations we have already given. For the series representation, we must modify (2) for nonintegral values of m, replacing $m!$ where it appears in that equation by the gamma function $\Gamma(m + 1)$, where the gamma function satisfies the functional relation

$$\Gamma(m + 1) = m\Gamma(m),$$

and where $\Gamma(\frac{1}{2}) = \sqrt{\pi}$. With this modification, (2) and (3) lead to Eq. (2.4), Chap. XII. The asymptotic behavior for large z given in Eq. (2.3), Chap. XII, follows directly from (4). To prove the trigonometric nature of the functions, we can first find $j_0(z)$ and $n_0(z)$ directly, showing that as given in (2.2), Chap. XII, they are given by $(\sin x)/x$ and $(\cos x)/x$, respectively. To do this, we can substitute the corresponding functions $J_{1/2}(z)$ and $N_{1/2}(z)$, defined from them by (2.1), Chap. XII, directly in Bessel's equation (1), and show that it is satisfied. Then we can derive the functions of higher index from these by the relation

$$\frac{d}{dz}[z^{-m}Z_m(z)] = -z^{-m}Z_{m+1}(z),$$

which we have given above, leading immediately to the algebraic and trigonometric nature of the higher functions, as well as to the explicit values for the functions.

SUGGESTED REFERENCES

Electromagnetic theory is too large a subject to be treated completely in a single volume, and we give a few references in the present section, which the student familiar with this book can refer to, without too great difficulty. In the first place, the reader of inadequate preparation may wish to review his elementary electromagnetism and optics, and can use, for example, *Introduction to Electricity and Optics*, by N.H. Frank (McGraw-Hill).

For general mathematical training and background, the student will first want texts on advanced calculus, such as *Treatise on Advanced Calculus*, by P. Franklin (Dover), or *Advanced Calculus*, by E.B. Wilson (Ginn). More advanced texts on analysis will be helpful, such as *Mathematical Analysis*, by Goursat and Hedrick (Ginn), *Partielle Differentialgleichungen der Physik*, by Riemann and Weber (Rosenberg), or *Higher Mathematics*, by R.S. Burington and C.C. Torrance (McGraw-Hill). Several texts on mathematics for special purposes are valuable: *Applied Mathematics for Engineers and Physicists*, by L.A. Pipes (McGraw-Hill); *The Mathematics of Physics and Chemistry*, by H. Margenau and G.M. Murphy (Van Nostrand); and *Mathematical Methods for Engineering*, by T. von Kármán and M.A. Biot (McGraw-Hill).

Among more specialized mathematical texts in the fields of particular importance to electromagnetism are *Introduction to Higher Algebra*, by M. Bôcher (Dover); *Ordinary Differential Equations*, by E.L. Ince (Dover); *Fourier Series and Spherical Harmonics*, by W.E. Byerly (Ginn); *Newtonian Potential Function*, by B.O. Peirce (Ginn); *Fourier Series and Boundary Value Problems*, by R.V. Churchill (McGraw-Hill); *Vector Analysis*, by H.B. Phillips (Wiley); and *Vector and Tensor Analysis*, by H.V. Craig (McGraw-Hill). The standard volumes of tables, *A Short Table of Integrals*, by B.O. Peirce (Ginn) and *Funktionentafeln*, by Jahnke and Emde (Dover), will be found invaluable for detailed assistance in calculation. For definite integrals that are not given in these books, *Nouvelles Tables d'Intégrales Définies*, by Bierens de Haan (Stechert), will be found a source of much information.

A number of general texts on theoretical physics cover electromagnetic theory among other topics. Among these we may mention

Introduction to Theoretical Physics, by L. Page (Van Nostrand); *Theoretical Physics,* by G. Joos (Stechert); *Introduction to Theoretical Physics,* by A. Haas (Constable); *Introduction to Mathematical Physics,* by R.A. Houstoun (Longmans); and *Principles of Mathematical Physics,* by W.V. Houston (McGraw-Hill). Two longer treatises on theoretical physics, in several volumes, may also be mentioned: *Introduction to Theoretical Physics,* by M. Planck (Macmillan), an English translation of a well-known German text, and *Einführung in die theoretische Physik,* by C. Schaefer (De Gruyter). The last two works go a good deal more into detail than is possible in the present book.

Next we come to a number of references dealing with the various branches of electromagnetic theory. Among general references, two standard works are *Classical Electricity and Magnetism,* by Abraham and Becker (Blackie), and *The Mathematical Theory of Electricity and Magnetism,* by J.H. Jeans (Cambridge). Treatments on approximately the scale of the present text are given in *Principles of Electricity,* by L. Page and N.I. Adams (Van Nostrand), *Electric Oscillations and Electric Waves,* by G.W. Pierce (McGraw-Hill), *Principles of Electricity and Electromagnetism,* by G.P. Harnwell (McGraw-Hill), and *Static and Dynamic Electricity,* by W.R. Smythe (McGraw-Hill). More advanced points of view are presented in *Electromagnetic Theory,* by J.A. Stratton (McGraw-Hill), and *Electromagnetic Waves,* by S.A. Schelkunoff (Van Nostrand)

A number of texts handle more specialized problems. In electron theory, and the theory of dielectrics and magnetic materials, there are *Theory of Electrons,* by H.A. Lorentz (Dover); *Theory of Electric and Magnetic Susceptibilities,* by J.H. Van Vleck (Oxford); *Polar Molecules,* by P. Debye (Dover); and *Introduction to Ferromagnetism,* by F. Bitter (McGraw-Hill). The theory of wave guides and microwaves is handled in a number of recent texts, including *Microwave Transmission,* by J.C. Slater (Dover); *Fields and Waves in Modern Radio,* by S. Ramo and J.R. Whinnery (Wiley); and *Hyper and Ultrahigh Frequency Engineering,* by R.I. Sarbacher and W.A. Edson (Wiley). The subject of optics is handled from the standpoint of electromagnetic theory in *Lehrbuch der Optik,* by K. Försterling (S. Hirzel), and *Optik,* by M. Born (Springer). Other standard treatments of optics are given in *Fundamentals of Physical Optics,* by F.A. Jenkins and H.E. White (McGraw-Hill), and in *Physical Optics,* by R.W. Wood (Dover).

INDEX

A

Absorption coefficient, optical, 108–112, 125–126

Ampère, 3, 53

Ampère, unit, 213

Ampère's law, 59–62, 83–85

Anomalous dispersion (*see* Dispersion)

Associated Legendre polynomials, or associated spherical harmonics, 32–34, 152, 224–226

Atomic refractivity, 116

B

Bessel's equation and function, 140–142, 152–156, 232–234

Biot-Savart law, 53–57, 63–64, 157–158

Boundary conditions, electromagnetic field, 44–46, 71, 117, 137

C

Capacity, 21–23, 206–207

Cavities, field in, 50–51, 75

resonant, 145–146

Charge density, 23

Clausius-Mosotti law, 110

Coefficients of Fourier series, 30–31, 164–165, 217–220

Coherence of light, 161–165

Condenser, 21–23, 46–47

energy of, 96

Conduction of electricity in metals, electron theory of, 106, 111–112

Conductivity, specific electrical, 85, 111–112

Conductor of electricity as an equi-potential, 15

Constitutive equations, 85

Continuity, equation of, 84

Continuity conditions, electromagnetic wave, 117, 137

Convergence of Fourier series, 217–220

Coordinates, curvilinear, vector operations in, 221–223

Cornu's spiral, 185–186

Coulomb, 10–12, 214

Coulomb's law, 2, 12, 158

Cross section, scattering, 160

Curl, in curvilinear coordinates, 222–223

of electric field, 14, 79–80, 83

of a vector, 14, 199

Current, density of, 61

displacement, 83–85

surface, 67

Curvilinear coordinates, vector operations in, 221–223

Cutoff in wave guide, 129, 146

D

D'Alembert's equation, 88

solution of, 169–173

Demagnetizing factor, 73–75

Density, charge, 23

current, 61

of energy in electromagnetic field, 95–103, 213

Diamagnetism, 65

Dielectric constant, 43–44, 105–114

relation of to polarization, 43–44

Dielectrics, 41–52, 105–114

types of, 41–42

Diffraction, 175–192

Dipole and dipole moment, electric, 35–42, 227–231

magnetic, 59, 70

oscillating, 157–161

Discontinuity in electromagnetic field, 44–46, 71, 117, 137

Dispersion, of electromagnetic waves in metals, 111–112

electron theory of, 105–114

Displacement, electric, 42–43

Displacement current, 83–85, 211